Business Earth Stations for Telecommunica
Walter L. Morgan and Denis Rouffet

Wireless Information Networks
Kaveh Pahlavan and Allen H. Levesque

Satellite Communications: The First Quarter Century of Service
David W. E. Rees

Fundamentals of Telecommunication Networks
Tarek N. Saadawi, Mostafa Ammar, with Ahmed El Hakeem

Meteor Burst Communications: Theory and Practice
Donald L. Schilling, Editor

Vector Space Projections: A Numerical Approach to Signal and Image Processing, Neural Nets, and Optics
Henry Stark and Yongyi Yang

Signaling in Telecommunication Networks
John G. van Bosse

Telecommunication Circuit Design
Patrick D. van der Puije

Worldwide Telecommunications Guide for the Business Manager
Walter H. Vignault

ADSL, VDSL, and Multicarrier Modulation

ADSL, VDSL, and Multicarrier Modulation

John A. C. Bingham
Palo Alto, California

A Wiley-Interscience Publication
JOHN WILEY & SONS, INC.
New York • Chichester • Weinheim • Brisbane • Singapore • Toronto

For ordering and customer service, call 1-800-CALL-WILEY.

Library of Congress Cataloging-in-Publication Data:

Bingham, John A. C.
    ADSL, VDSL, and multicarrier modulation / John A. C. Bingham.
      p.   cm.—(Wiley series in telecommunications and signal
  processing)
    "A Wiley-Interscience publication."
    ISBN 0-471-29099-8 (alk. paper)
    1. Digital telephone systems.   2. Telecommunication—Standards.
  3. Modulation (Electronics).   4. Computer networks.   I. Title.
  II. Series.
  TK5103.7.B535   2000   99-15963
  621.385—dc21

Printed in the United States of America.

10 9 8 7 6 5 4 3 2 1

To my dear wife, Lu

CONTENTS

PREFACE

I reread the preface of my first book [Bingham, 1988] and was very tempted to reproduce much of it here. The style and intended audience of the two books are much the same: both are something between an academic textbook and an engineering handbook and are aimed primarily at design engineers and programmers. The level of mathematics assumed is, for the most part, about first-year postgraduate, with only occasional excursions into more exotic realms.

The *and* in *ADSL, VDSL, and Multicarrier Modulation* is not precise; the scope of the book is wider than the intersection (a logic designer's *and*) but narrower than the union (a layperson's *and*). On the one side there are some types of multicarrier modulation (MCM) and some applications of it that are not covered, and on the other side some modems for the digital subscriber line (generically called xDSL) that are not covered; I have tried to provide enough references to take an interested reader further in those subjects.

The intersection—MCM used for the DSL—is a hot topic right now. Discrete multitone (DMT) has been standardized for asymmetric DSL (ADSL) by the American National Standards Institute (ANSI) as T1.413 and by the International Telecommunications Union (ITU) as Recommendation G.992 and may soon be standardized for very-high-speed DSL (VDSL). My hope, however, is that some of the material in this book will be general and forward-looking enough that it can be used—long after the glare of "Internet access" publicity has faded—to spur improvements in ADSL and VDSL.

These improvements should, as in all telecommunications, be backward compatible with previous-generation systems. Such compatibility will, however, be more difficult for DMT and ADSL because DMT was chosen and defined as a standard before the technology was mature. DMT is like the pianist Van Cliburn: heaped with honors early in its career and in danger of being chained to a metaphorical Tschaikovsky's Piano Concerto forevermore. The developers of DMT in the next few years could confine themselves to the receivers—thereby avoiding any problem of backward compatibility—but this would limit their creativity too severely. A better strategy (and a bigger challenge) is to develop better transmitters that are not so different from the standardized ones that they cannot economically be included as options, and are activated only when connected to a compatible unit. G.994.1 defines an etiquette[1] for "handshaking" during the initialization of ADSL modems, which should allow for such future developments.

I have many ideas about these improvements, but since I am retiring I will not be able to work them out. I have therefore suggested them, and then used the term *unfinished business*. It is important to realize, however, that these improvements will not bring the increase in data rates that have been achieved recently in voice-band modems: a factor of 2 approximately every six years for the last 20 years or so. Despite their immaturity, DMT ADSL modems are probably operating within about 5 dB of the performance that is theoretically achievable under near-worst-case noise conditions. Improvements will come in the ability to deal with—usually to take advantage of—the widely varying levels of noise that occur in practice and in the practical matters of cost, size, and power.

During the discussions leading up to the adoption of the DMT-based standard there was intense intellectual and commercial rivalry between MCM and the more classical single-carrier modulation (SCM) methods. This rivalry, in which I enthusiastically participated, had the effect of discouraging—and in many cases preventing—objective discussion of the relative merits of the methods. I am retired now and can be a little less biased, but am probably still not yet far enough removed to write a completely objective comparison; therefore, I will try just to describe MCM, and mention SCM only when similarities or differences help to explain MCM.[2] The reader is referred to [Saltzberg, 1998] for an excellent comparison of SCM and the immature DMT as it existed in 1998. Whether his assessment of the relative advantages of the two methods will be valid as DMT matures remains to be seen.

One of the factors in the commercial and intellectual competition is the intellectual property (IP) owned by the competing companies, and patents are an important part of every engineer's library. I will therefore list all relevant patents that I know of, but I must make an emphatic disclaimer that I hope readers will empathize with: citing a patent means only that I consider that the idea has technical merit; it implies no opinion about the patent's legal validity.

DMT for ADSL was first developed at Amati, and was so successful that TI bought us in 1998. There was a rumor for a while[3] that in recognition of our contribution they would change their name to California Instruments, but alas, it was Amati's name that changed: AmaTI, then AmaTI, and now just TI[4]!

I am very pleased to have three contributors to this book: one collaborator on the T1E1.4 committee, Alan Weissberger, one ex-colleague, Mitra Nasserbakht,

---

[1] See [Krechmer, 1996] for a discussion of etiquettes and protocols as they operate in the world of standards.
[2] I will probably not be able to resist a chauvinistic comment from time to time, but I will try to confine them to the footnotes.
[3] I confess; I started it on April 1, 1998!
[4] The Amati family were the first makers of really good violins. There is no evidence that Stradivarius bought out Amati, but otherwise there is a close match.

and one group of ex-competitors from Aware Inc. They are experts in ATM, FFT implementation, and DWMT, respectively, and essential contributors to the overall MCM picture.

ACKNOWLEDGMENTS

I am much indebted to Amati Communications and particularly to its founder, John Cioffi. John is a good friend, a brilliant engineer, and was a provocative and inspiring leader. I thank him and everybody at Amati for the most exciting and rewarding last six years of a career that any engineer could hope for.

I am indebted to my colleagues on the T1E1.4 committee who wrote the ADSL standard, and especially to Tom Starr, the exemplary chairman of that committee. I am also indebted to Jean Armstrong, Gianfranco Cariolara, Donald Chaffee, Jackie Chow, Peter Chow, John Cook, David Forney, Kevin Foster, Hans Frizlen, Umran Inan, Krista Jacobsen, Anjali Joshi, Jack Kurzweil, Phil Kyees, Joe Lechleider, Masoud Mostafavi, Joseph Musson, Dennis Rauschenberg, Craig Valenti, Joe Walling, Brian Wiese, Kate Wilson, and George Zimmerman for many helpful discussions.

ADSL, VDSL, and Multicarrier Modulation

1

INTRODUCTION

The four principal media for transmission of high-speed data to and from a customer premises are:

1. *Subscriber telephone loop [digital subscriber loop (DSL)]:* the unshielded twisted pair (UTP) of copper wires used for "plain old telephone service" (POTS)
2. *Coaxial cable:* originally installed for unidirectional ("downstream") transmission of television, but increasingly being used for bidirectional data transmission
3. *Optical fiber:* originally used for very high-speed trunk transmission, but now being considered for either the last leg of the distribution [fiber to the home (FTTH)] or the penultimate leg [fiber to the exchange or fiber to the neighborhood (FTTE or FTTN)]. The latter case is the only one that will concern us, because then the last leg is provided by the distribution portion of the DSL (see Section 3.1).
4. *Wireless.*

There is no general answer to the question of which of these is *best*, and the four have contended vigorously for many years for both media attention and developmental and deployment capital. In this book we are not concerned with the rival merits—technical, financial, political, social, or environmental—of these four media[1]; we will describe only the first. We are concerned only with the physical layer (the lowest layer) of the OSI model; in Chapter 2 we deal with the upper part of that layer—the transmission convergence (TC) layer—and in the rest of the book, with the lower part—the physical medium-dependent (PMD) layer. The main topic at the PMD level is multicarrier modulation (MCM)— in particular, discrete multitone (DMT)—applied to xDSL. There are, however, many types of DSL (e.g., ISDN, HDSL, SDSL) that do not use MCM, and furthermore, MCM is used in media (particularly wireless) other than DSL; we

---

[1] The perception of the merits seems to have depended on who put out the last set of press releases!

1

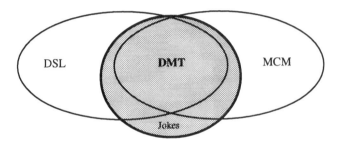

Figure 1.1 Scope of this book.

discuss these only briefly as an introduction to the main topic. This is illustrated in the Venn diagram of Figure 1.1; the scope of the book is less than the union but greater than the intersection.

1.1 ARRANGEMENT OF THIS BOOK

In the remainder of this chapter we describe, in sequence, the histories of DSL, MCM, and MCM applied to xDSL. In Chapter 2, by Alan Weissberger (which probably could be expanded to be a book by itself), the TC layer is discussed. In Chapter 3 we describe the physical medium, and in Chapter 4, ways of using the medium for data. There is no completely logical order or grouping of topics thereafter. In Chapters 5, 6, and 7 the theory of MCM is discussed: the fundamentals in Chapter 5; discrete multitone (DMT), a simple version of MCM, in Chapter 6; and general MCM in Chapter 7. Chapters 8, 9, and 10 are practical, dealing with the implementation of DMT as ADSL and VDSL. Chapter 11 is the "fun" one: a discussion of some possible future improvements for A, V, and xDSL in general.

1.2 HISTORY (ONGOING) OF DATA ON THE DSL

It is difficult to say when the subscriber loop was first used for data (telegraph; 110-bit/s voice-band modems?), but the systems that are still around are as follows.

- *Basic rate access DSL* (also known as just DSL). 160-kbit/s one-pair full-duplex system. Used in the United States only for data services to provide access to the Integrated Services Digital Network (ISDN) [ANSI, 1993b], but also used in Europe for 2×64 kbit/s digitized voice service. ITU Recommendation G.961 defines three different systems:
  - *Appendix I:* 2B1Q coding with echo cancellation (EC); used in North America and much of Europe. Also standardized in North America as T1.601: see [ANSI,1993].

- *Appendix II:* 4B3T coding with EC; used in some European countries.
- *Appendix III:* bipolar (a.k.a. AMI) coding with synchronized time-division duplexing (TDD; a.k.a. "ping-pong"); used in Japan.
- *T1.* 1.544-Mbit/s dual simplex on two pairs using AMI coding and repeaters spaced every 6 kilofeet (kft); used in North America. T1 was originally designed, and installed [Cravis and Crater, 1963] from 1962 onward for interoffice (trunk) transmission of 24 multiplexed 64-kbit/s PCM voice channels; for that use it has now been almost completely replaced by fiber and microwave. Since the early 1970s, however, it has also been used on the DSL, and it is by far the most severe potential source of crosstalk into ADSL.[2] It will be made obsolete by HDSL2, but it is very unlikely that installed systems will be replaced.
- *E1.* Similar to T1, but 2.048 Mbit/s for 32 voice channels, with repeaters spaced approximately every 2 km; used everywhere else in the world.
- *High-speed DSL (HDSL).* 1.536-Mbit/s two-pair and 2.048 Mbit/s two- and three-pair, full-duplex systems using 2B1Q coding and echo cancellation: originally defined in [ANSI,1994] and [ETSI,1995], and now codified as ITU Recommendation G.991.1.
- *Asymmetric DSL (ADSL).* ANSI standard T1.413 [ANSI, 1995] defines an ADSL system to transmit downstream and upstream data rates up to 6.8 and 0.64 Mbit/s, respectively, within a radius of approximately 12 kft from the CO [known as the carrier serving area (CSA)], and 1.544 and 0.176 Mbit/s within a radius of 18 kft [the extended CSA (ECSA)]. ITU Recommendation G.992.1 defines a system based on T1.413 as a core, but expanded via three annexes to meet particular regional needs. G.992.2 defines a simpler system with a wider range of data rates and ranges (see Section 1.5 on ADSL lite) that is line compatible with G.992.1. ADSL is the main subject of this book, and T1.413 and/or G.992 should be indispensable companions while reading.
- *Very high-speed DSL (VDSL).* VDSL will be used primarily in "hybrid fiber/copper" systems to connect optical network units (ONUs) to customer premises. In fiber to the exchange (FTTE) systems these ONUs will be in the CO, and we will call the VDSL transceivers VTU-Cs. In other systems—FTTN(eighborhood), FTTC(urb), and FTTB(uilding)—the ONUs will be outside the CO; the only difference between these will be the length of the loop from ONU to the customer premises: up to 6 kft for FTTN or 1.5 kft for FTTB systems. We will call them all FTTC(abinet) systems, and the transceivers VTU-Os. If the location (CO or outside ONU) is not important for a particular discussion we will call the "head-end" transceiver VTU-C/O. VDSL ranges vary from 1 to 6 kft, depending on the location of the ONU, and corresponding aggregate (down plus up) data rates vary from approximately 58 to 4.6 Mbit/s. Two

[2] See Section 4.5 for a discussion of this.

modes are defined in [Cioffi, 1998]: asymmetric with a down/up ratio of approximately 8/1, and symmetric. Three line codes have been proposed: DMT, Zipper (a variant of DMT), and CAP (a variant of QAM).

- *HDSL2.* 1.536-Mbit/s one-pair full-duplex system using a mixture of frequency-division duplexing and echo cancellation, and very sophisticated trellis coding. Probably will be standardized by ANSI in 1999 and by the ITU as G.991.2.
- *SDSL.* Various unstandardized one-pair full-duplex systems achieving less than 1.536-Mbit/s. The advantages over HDSL2 may include lower cost, earlier availability, and greater range.

The general pattern has been for each successive system to use a wider bandwidth than the preceding one, and a totally different, non-backward-compatible modulation scheme.

1.3 HISTORY OF MULTICARRIER MODULATION

The principle of transmitting a stream of data by dividing it into several parallel streams and using each to modulate a "subcarrier" was originally applied in Collins' Kineplex system,[3] described in [Doelz et al., 1957]. It has since been called by many names, and used—with varying degrees of success—in many different media:

- *FDM telephony group-band modems.* [Hirosaki et al., 1986] described an *orthogonally multiplexed QAM* modem for the group band at 60 to 108 kHz. It used a fixed bit loading (see Section 5.3), and its main advantage over single-carrier modems was a much reduced sensitivity to impulse noise. I do not know if there are any still deployed.
- *Telephony voice-band modems.* [Keasler and Bitzer, 1980] described a modem for use on the switched telephone network (STN), and in 1983 Telebit Corporation introduced the Trailblazer modem [Fegreus, 1986], which used *dynamically assigned multiple QAM*. It far outperformed all single-carrier contemporaries, and for certain applications (e.g., file transfer using UNIX) it was ideal. It was proposed as a standard for an STN modem [Telebit, 1990] but was rejected because of its very large latency.[4]
- *Upstream cable modem.* [Jacobsen et al., 1995] proposed *synchronized discrete multitone* (SDMT) for the 5- to 40-MHz upstream band in a hybrid fiber coax (HFC) system. SDMT uses a combination of frequency-

---

[3] I did hear a claim that there was a system before Kineplex, but I do not remember the details. If there was such a system, I apologize to the developers for slighting them.

[4] It used 1024 subcarriers with a spacing of approximately 4 Hz.

division multiple access (FDMA) and time DMA (TDMA) and is ideally suited to both the medium and the system requirements, but it faded because of lack of commitment and a sponsor. I do not know whether it is now dead or just cryogenically preserved. The name SDMT is now used to describe another synchronized version of DMT proposed for VDSL.

- *Digital audio broadcasting. Coded orthogonal frequency-division multiplexing* (COFDM)[5] is a version of MCM that uses IFFT modulation (see Section 6.1), fixed bit loading,[6] and sophisticated coding schemes to overcome the fades that result from multipath. It has been standardized in Europe as the Eureka system [OFDM1].

- *Digital audio radio.* A version of DMT for use in the United States in the same frequency bands as the established FM stations was tested in 1994. It performed as well as could be expected in the very severe narrowband, low-power, high-noise (from the FM signal) multipath-distorted environment, but that was not good enough for widespread deployment. In-band digital radio is currently on the back burner in the United States.

- *Digital TV.* COFDM has also been standardized for digital video broadcasting [OFDM2].

The subtitle of [Bingham, 1990] was "An idea whose time has come" but "has come" at that time clearly should have been "is coming", "may come", "came and went", or "probably will never come", depending on what application and/ or transmission medium was being considered.

Other Forms of MCM. All of the foregoing systems used sinusoidal subcarriers, but a more general form of MCM, which uses more complex signals as "subcarriers" in order to maintain orthogonality in a distorted channel was originally proposed in [Holsinger, 1964]; it has since had many different forms, which are discussed in Chapter 7.

1.4 MCM (DMT) AND DSL

The use of DMT for ADSL was first proposed in [Cioffi, 1991]. In 1992, ANSI committee T1E1.4 began work toward a standard for ADSL, defined a set of requirements, and scheduled a competitive test of all candidate systems. The tests were performed on laboratory prototypes in February 1993, and in March 1993 the DMT system was chosen to be the basis of the standard. I took over as editor of the standard in 1994.

Representatives of all seven regional bell operating companies (RBOCs), most European national telcos (previously, PTTs), and at least 30 telecommu-

---

[5] See the specialized bibliography in the reference section.
[6] In a broadcast mode there can be no feedback from receiver to transmitter.

nications manufacturers from throughout the world participated in the drafting and revising process, and in August 1995, Issue 1 of ANSI Standard T1.413 was published. As is usual with such standards, changes were suggested at the last minute that were too late to be included in Issue 1, and work was started immediately on Issue 2. This work proceeded rather desultorily, however, because market demands had changed since the original project was defined. 6 + Mbits/s downstream for high-quality compressed video ("video on demand") no longer seemed economically attractive, and there was a danger that T1.413 would become a standard without an application.

Then in early 1996 access to the Internet became paramount. As [Maxwell, 1996] put it, "... simply uttering the word Internet before securities analysts doubled a company's stock price." ADSL was reborn with a different persona:

- 6 + Mbit/s to perhaps 50% of all households became less important than 1.5 Mbit/s to perhaps 80%.
- ATM became a much more important transport class of data than STM.
- Dynamic rate adaptation—the ability to change data rates as line conditions (mainly crosstalk) change—became important.

Work was redirected accordingly, and Issue 2 was published early in 1999.

ITU Study Group 15 began work on xDSL in late 1997 and addressed the questions of unique national and regional needs (see Appendix B.1). G.992 for ADSL was published in 1999.

1.5 ADSL "LITE"

T1.413 was still, however, perceived by many—particularly those in the computer industry—as being too complicated, expensive, and telco-centric. This prompted demand for a "lite" modem. SG 15 took over responsibility for what was temporarily called *G.lite* and is now designated G.992.2. The characteristics—some fairly precise, some rather vague—of a G.lite modem were billed as:

1. User-friendly; that is, very few options, take it out of the box, plug it in without requiring assistance from the phone company,[7] and use it.
2. Less complex; therefore, presumably, less expensive.
3. No rewiring of customer premises should be needed; existing house wiring, no matter how ancient and chaotic, should be adequate.
4. The low-pass part of the POTS splitter (see Section 9.1) should not be needed.
5. Only transport of ATM should be supported.

---

[7] No "truck roll."

6. Range should be the more important than rate; some service, albeit at only $0.7 + \text{Mbit/s}$ downstream, should be possible out to 22 kft.
7. "Always on"; that is, an ATU-R should have a standby mode in which it would use very little power, but be ready—within some small-but-still-to-be-defined time—to receive email and other unsolicited downstream transmissions.

Requirement 4 started out as the most important, but was modified as work progressed.

1.6 SOME HOUSEKEEPING DETAILS

1.6.1 Units of Measurement

In most scientific and engineering books there would be no question that the metric system of measurement should be used exclusively. In discussing telephone systems, however, the issue is not as clear. In the United States, wire sizes and lengths are measured in American wire gauge and—in a strange, halfhearted attempt at metrification—kilofeet, and most of my experience has been in those units. Therefore, I will use them primarily and, wherever appropriate, show conversions to the metric system. I will use the compatible set of units: kΩ, nF, mH, and MHz in all except one case: dBm/Hz is too firmly entrenched to be dislodged by the more convenient dBm/MHz.[8]

1.6.2 References

In order to help readers recognize references without having continually to flip to the end of the book, we cite them as [Smith and Jones, 19xy] without worrying about whether we are referring to the paper or the authors. On some topics we have included block bibliographies at the end of the reference section without citation or recommendation of any particular paper.

---

[8] Both of them are, of course, mathematically inconsistent (x dBm/Hz does not mean $2x$ dBm in 2 Hz!), but mW/Hz never caught on.

2

ADSL NETWORK ARCHITECTURE, PROTOCOLS, AND EQUIPMENT

A. J. Weissberger

P.O. Box 3441, Santa Clara, CA 95055-3441, E-mail: alan@lambdanetworks.com

2.1 ADSL ADVANTAGES AND APPLICATIONS

ADSL is attractive to both telcos and users, because it solves two problems simultaneously:

1. It provides a simple, affordable mechanism to get more bandwidth to end users: both residential and small- to medium-size business. This is increasingly important for Internet access, remote access to corporate servers, integrated voice/data access, and transparent LAN interconnection.

2. It enables carriers to offer value-added, high-speed networking services, without massive capital outlays, by "leveraging" the copper loop. Examples include access to frame relay or ATM networks, virtual private networks, video distribution, streaming, or video retrieval services.

In North America, the driving applications of ADSL are high-speed Internet access and remote access to corporate LANs. Other applications include video retrieval or streaming, interactive multimedia communications, video on demand, video catalog shopping, and digital telephony: either voice telephony over ATM or voice over IP (VToA and VoIP).

In Asia and parts of Europe (e.g., United Kingdom and Germany) video on demand (VoD) and audio playback are much more important than in the United States. Ironically, VoD was the ADSL application driver for North America in 1993 but has since been abandoned by most U.S. telcos. In Asia, however, where there is not as large an installed cable TV customer base, ADSL could be very important for video and audio distribution.

2.2 ADSL TRANSPORT MODES: STM OR ATM?

The original ADSL standard was designed to carry compressed digital video (i.e., MPEG2), $n \times 64$ kbit/s and DS1 dedicated circuits. This class of information transfer is known as *synchronous transport mode* (STM). With the redirection of ADSL to transport IP packets, there was a movement to support variable-length frames (e.g., HDLC or Ethernet MAC) as part of STM. Since 1997 ATM, or cell-based transport, has been favored over STM (in order to support IP packets as well as compressed video and other real-time or QOS-based applications), and G.992.2 (G.lite) supports *only* ATM transport.

Since the majority of telco networks now have ATM backbones, the extension of ATM over the subscriber enables the telco to take advantage of economies of scale. It also dispenses with protocol conversion at the access-network-to-core-network-interface. Finally, an ATM network can more easily scale up to accommodate more subscribers and/or higher access speeds. This would make it easy for a carrier to accommodate growth in both the numbers and downstream bit rates of ADSL lines and to build the infrastructure for VDSL (see the discussion on network architecture in Section 2.3).

With ATM over ADSL, users are connected to a network service provider (NSP) via *virtual circuits*[1]. Currently, both a PPP over ATM stack (for Internet and secure corporate server access) and a native-mode ATM protocol stack (for real-time and multimedia applications) are used in conjunction with PVCs. In the future ATM SVC signaling (a.k.a. ATM Forum UNI or ITU Q.2931 signaling) and ATM network management (ATM Forum ILMI) messages will be supported in the access node and the ATM over ADSL CPE.

For ATM over ADSL as defined in T1.413 or G.992.2, user data is segmented into cells, which are then transmitted and received over the subscriber loop by the pair of ADSL modems (the NT on the customer premises and the access node in the network—typically on a line card within a DSLAM or ATM edge switch).

The ATM network supports various *traffic classes* to realize the desired user service. These are specified on a virtual circuit basis, along with subordinate traffic class/QOS parameters. From highest to lowest priority, these traffic classes are:

1. Constant bit rate (CBR)
2. Real-time and non-real-time variable bit rate (VBR)
3. Available bit rate (ABR)
4. Unspecified bit rate (UBR)

---

[1] Today these are *private virtual circuits* (PVCs), but carriers plan to offer *switched virtual circuits* (SVCs) in the future. In the meantime two techniques—*soft PVCs*, which are effectively PVCs that have been set up but never taken down, and *auto-configuration extensions* to the ILMI MIB—can be used for more flexible provisioning.

Today, most ADSL networks use only UBR, but those supporting high-quality video or audio also use CBR. Those ADSL modems that support both these traffic classes must implement multiclass queuing and traffic scheduling, so as always to give priority to CBR traffic.

All three ADSL-DMT standards specify the same *cell TC* for mapping ATM cells into the user data field of an ADSL physical layer frame. There are separate cell TCs for the interleave and fast paths: corresponding to the ADSL channels (AS0 and AS1 downstream and LS1 and LS2 upstream) in use. Only one channel, in each direction of transmission, exists for G.992.2, but up to two upstream and downstream channels are optional in T1.413-II and G.992.1[2]. Hence for dual latency in a given direction of transmission, the cell TC appears as two physical layers to the ATM layer. An example of this would be the concurrent use of video retrieval on the interleave path and Internet access or digital telephony (e.g., VToA) on the fast path.

In addition to cell delineation, the cell TC performs other functions:

1. It inserts and removes idle cells from the ADSL physical layer user data.
2. It scrambles/descrambles the cell payload.
3. It checks for HEC violations on each received cell and discards cells with HEC errors.
4. It performs sublayer bit timing ordering.
5. It reports both the inability of the receiver to acquire cell delineation (no cell delineation) and the loss of cell delineation after it had been acquired. These anomalies are reported in *indicator* bits within the ADSL superframe.

The ATU-R is required to maintain three cell TC counters to monitor cell TC performance.

Sublayer interfaces for the cell TC are defined in a T1.413-II Annex for the ATM layer above (nominally, the UTOPIA or UTOPIA 2 interface from the ATM Forum) and the sync/control multiplexing PHY sublayer below. Again, one cell TC is required for each latency path/ADSL channel.

2.3 ATM END-TO-END NETWORK ARCHITECTURES AND PROTOCOL STACKS

Initially, ATM over ADSL modems used PVCs and IETF RFC 1483 bridging to encapsulate user data into AAL5 packets and then into ATM cells. The modems were transparent to the higher layer protocols (e.g., TCP/IP, IPX, Appletalk, etc). Each customer was preassigned a local label in the ATM cell header (VPI.VCI) to correspond to the NSP with which the customer wanted to

---

[2] Sometimes called G. regular to distinguish it from G.lite!

communicate. For example, one PVC could be assigned for an ISP and another to communicate with corporate headquarters. However, higher layer protocols used by NSPs could not effectively be overlaid on top of the RFC 1483–based protocol stack.

For Internet access and remote access to corporate servers it was highly desirable to use the same "legacy" programs for authentication, billing, and encryption/security that are used by NSPs today. These are all operational over the Internet engineering task force (IETF)'s *point-to-point protocol* (PPP). To facilitate the PPP over ATM over ADSL capability, the ADSL Forum has completed TR-0012: an end-to-end architecture for the transport of PPP over ATM over ADSL. This is likely to be implemented by the majority of ADSL equipment vendors. In this scheme, the entire ATM network is reduced to a set of virtual point-to-point leased lines, and all traffic is sent on a "best effort basis" using the ATM *unspecified bit rate* (UBR) traffic class.

Whereas today, only PVCs are used with ADSL, a catalyst for SVCs will be widespread use of Microsoft's ATM protocol stack, including SVC signaling, in Windows 98 and Windows 2000 (formerly known as Windows NT). This will encourage use of SVCs end to end, which are much easier than PVCs to maintain in a large network. A potential problem for SVCs is that the ATM address plans of telcos differ, creating nonunique ATM addresses, which may be either E.164 public network addresses or private *network service access points* (NSAPs). Another issue is mapping SVC UNI signaling messages to/from the ADSL facilities and the ATM network behind the access node (see the discussion of DSLAM in Section 2.3.1).

Since there is no *quality of service* (QoS) capability or multicasting with the PPP over ATM architecture, vendors desiring to provide video/high-quality audio on demand, VToA/VoIP, or real-time video conferencing over ADSL, must chose a *classical ATM protocol stack*. These stacks have been well defined by the ATM Forum and ITU-T and include:

- MPEG2 over ATM (using AAL 5)
- Structured circuit emulation service[3] (for $n \times 64$ kbit/s circuits)
- VToA desktop[3] (using AAL 5 or AAL2)
- Video conferencing over AAL 5 or AAL 1

Thus there are three ATM protocol stacks for ATM over ADSL:

- RFC 1483 encapsulation/bridging
- PPP over ATM as in ADSLF TR-0012
- Classical ATM (ATMF and ITU-T for real-time, interactive applications such as VoD, video streaming, VToA, conferencing, etc.)

---

[3] These applications may use the network timing reference (see Section 8.2.1).

2.3.1 New Equipment Needed for ADSL

In addition to the ADSL central site and remote site modems, the key new network infrastructure equipment required to make ADSL a commercial reality is the *digital subscriber-line access multiplexer* (DSLAM). This equipment aggregates a large number of ADSL subscribers into one or a few uplink ports to a frame relay or ATM backbone network (edge switch or router). The uplink interface is formally known as the V reference point. It is typically a DS3/E3 facility, but could also be $n \times$ DS1/E1 with inverse multiplexing, or even SONET OC3c/STM-1 (155 Mbit/s).

NOTE: The architecture of the DSLAM may determine the entire design of the ADSL network. Some DSLAMs handle only ATM over ADSL (e.g., Alcatel); others support a variety of DSLs with both ATM and frame-based transport (e.g., Ascend).

The DSLAM acts as a VPI/VCI cross-connect for PVCs (the VPI/VCI "labels" have only local significance for a particular point-to-point ATM link). It must aggregate traffic from ADSL links and map to uplink. Conversely, the DSLAM must distribute traffic from the uplink to the appropriate ADSL port (more than one, if multipoint virtual circuits are supported). UPC/policing of ATM traffic contracts will also be required in the DSLAM.

For SVCs, UNI signaling messages could be passed transparently through the DSLAM in configurations known as *virtual UNI* and *SVC tunneling*. However, there are major problems with these methods that will greatly restrict their use. More likely, the DSLAM will terminate UNI signaling messages (as the network side of the UNI) and map them over the V reference point, as either the user side of the UNI or an *access node-to-node interface* (ANNI). In this scenario, there are independent signaling state machines, each of which has intimate knowledge of the link(s) to which they are connected. Hence *connection admission control* (CAC) and QoS parameter negotiation can be done properly at call setup time.

With the great interest in PPP over ATM, there is a perceived need for new adjunct equipment (behind the DSLAM) for concentrating or terminating PPP sessions at the boundary between the network service provider (NSP) and the access network: for example, either an *L2 access concentrator* (LAC) or a *broadband access server* (BAS). The LAC concentrates multiple PPP sessions into a smaller number of PVCs to the NSP's broadband network. The BAS terminates the PPP sessions and probably handles authentication, billing, and security functions (if needed). The LAC and BAS may be colocated with the DSLAM in a central office, or many DSLAMs can be connected to a single LAC/BAS. It is expected that one or the other of these adjuncts will be deployed in either an access provider or ISP network.

The formal specification of these and other network equipment adjuncts will be done by the ADSL Forum under *core network architecture*.

NOTE: Since so much of the ADSL network architecture will be reusable by VDSL, it is imperative for ADSL to be successful if VDSL is to leverage off it. This includes DSLAMs, access multiplexers, ATM over ADSL NTs, standardized protocol stack and network management, and so on. Hopefully, the ATM access network being developed for ADSL will be fully operational and reliable by the time telcos are able to deploy VDSL in a big way. This will greatly increase VDSL's chances of success.

2.4 MAPPING DIGITAL INFORMATION TO ADSL USER DATA

2.4.1 Premises Architecture and DTE-to-DCE Interface

The method to map user data to ADSL PHY will depend on the customer premises configuration. This is likely to be one of the following:

- *Integrated network interface card (NIC):* especially for G.992.2
- *Single user (via bridging):* 10BaseT, *universal serial bus* (USB), ATM25
- *Multiple users (via routing):* twisted-pair and wireless home networks, 10 Base T, IEEE 1394

In the last two configurations a specific DTE-to-DCE interface (e.g., PC to ADSL NT) will be required. This interface is currently not standardized. However, there are several candidates for premises architectures. These include:

- *Broadband media access protocol* (BMAP) in ADSL Forum and USB Interoperability Group
- PPP over Ethernet (PPPOE) in ADSL Forum
- *Frame-based UNI* (FUNI) over Ethernet in ATM Forum
- *Layer 2 tunneling protocol* (L2TP), which is an IETF draft (note that PPTP is Microsoft's version of this)

BMAP and PPPOE extend the PPP session to the client PC, while FUNI is independent of PPP. Both BMAP and FUNI take advantage of an ATM protocol stack in the PC (e.g., from Microsoft), which effectively enables "ATM to the desktop" without an ATM premises PHY. PPPOE assumes an IEEE 802.3/Ethernet interface with no ATM stack in the PC.

When supporting PPP to the client PC, the ADSL NT must be able to map a PPP session to the associated virtual circuit label (i.e., VPI/VCI). Each of the premises architectures noted above has a different procedure to do that.

NOTE: It is interesting that none of the ATM bridging/routing specifications defined previously (e.g., LAN Emulation, MPOA, RFC 1577) is being seriously considered for ATM over ADSL.

2.4.2   Traffic Shaping

Once the bridging/routing technique is fixed within the ADSL NT, its next concern is *traffic shaping* of the ATM cells from the DTE to the US ADSL channel (LS0 or LS1). This will prevent user-generated data, arriving at 10 Mbit/s over a 10 Base T link, from exceeding the ADSL US channel rate of perhaps 384 kbit/s. Traffic shaping smooths out cell transmissions so as not to exceed a predefined *peak cell rate* (PCR) for a given virtual circuit. The sum of all active PCRs should not greatly exceed the PHY layer bandwidth; otherwise, cells will be lost during busy traffic periods. While traffic shaping for UBR class is optional in ATM Forum and ITU-T specifications, it will be mandatory for ADSL according to ADSL Forum WT-21[4] (revision to TR-0002).

2.4.3   Single or Dual Latency at the ATM Layer

If all ADSL communications are over a single ADSL channel (e.g., G.992.2 using the interleave path), then each ATM endpoint (e.g., ISP, corporate HQ, partner company site, etc.) has an ATM address (for SVCs) or preassigned VPI-VCI for each PVC. The ambiguity comes in an SVC when there are two ADSL channels (fast and interleave paths). Where is it decided to which latency path (i.e., ADSL channel) the requested SVC should be assigned? Note that because SVCs are set up and cleared dynamically, this information cannot be provisioned! Remember that the latency path is chosen independently for each direction of transmission. Also, the ADSL channels for ATM (AS0, AS1 DS and LS0, LS1 US) are unidirectional (simplex) and correspond one-to-one to the latency path.

For dual latency downstream, an "intelligent" ADSL access node (usually in a DSLAM) may be able to select a latency path based on QOS parameters/information elements in the setup message, (e.g., CLR, CTD, CDV, etc.). The ADSL NT or DTE terminating UNI signaling messages would simply accept that VPI/VCI mapping to the latency path selected. Many telcos originally thought that dual latency would be needed only for downstream (e.g., for motion video on interleave and real-time applications or Internet/Intranet access on fast path). In this case, the ADSL facility would be configured for dual latency downstream and single latency upstream. Some telcos, however, are now saying they would like to have dual latency upstream as well: for burst-error-protected Internet/Intranet and SVC signaling on the interleave path and real-time applications (VoIP or VToA or video conferencing) on the fast path. Note that there is only a single size for the interleave buffer for all VPI/VCIs that use that path. The buffer depth is chosen to be commensurate with the maximum impulse noise burst expected. It may be on the order of 40 or 50 ms per ADSL link. Therefore, there needs to be a new mechanism to specify the latency path to VPI/VCI mapping for SVCs. Whether this is to be done by a new information element in

---

[4] See Appendix B.5 for information on ADSL Forum documents.

the UNI signaling message or by convention (e.g., odd VPI/VCI for fast path; even for interleave path) has yet to be determined by the ADSL or ATM Forums. Once the latency path mapping has been determined, a new ATM layer function must assign each cell to be transmitted to the designated latency path. The VPI/VCI in each cell becomes an index to a 1-bit lookup table that specifies the correct path (ADSL channel).[5] In addition to selecting a latency path for user data, one must also be assigned for both signaling and ILMI (ATM access via SNMP) messages. If we have dual latency downstream and single latency upstream, which downstream path should be selected for these control and management protocols? This assignment has yet to be standardized.

In a dual latency environment, the ability to reassign bandwidth from one latency path to another after modem startup is known as *rate repartitioning* (RR). Since bandwidth usage is not static, RR would be highly desirable. It is optional in T1.413-II, and specified in informative Annex K. However, the means for the ATM layer to request RR and notify of its completion has yet to be standardized (see WT-21 open issues in Section 2.6).

2.5 UNIQUE ADSL REQUIREMENTS FOR ATM

Many of the ATM over ADSL issues are addressed in ADSL Forum WT-21 [ADSLF, 1998]:

- Specific reference models: functional blocks and interfaces
- Transport of ATM over ADSL, including the problems presented by dynamic rate change:
  - DRA and RR (dual latency mode) for full-rate ADSL (T1.413 and G.992.1)
  - Fast retrains and full retrains/restart for G.992.2
- QOS and traffic management
- Functional block definitions
- ATM Forum and ITU-T signaling (for SVCs)
- Management: including use of OAM cells according to ITU Recommendation I.610
- ATM virtual circuit assignment
- Annexes on:
  - Relationship to other reference models
  - Standards work cross reference
  - SVC call load analysis
  - ATM VP/VC assignment in dual latency mode

---

[5] None of the commercially available ATM SAR chips has this capability today; they will need to be modified for dual latency full-rate ADSL.

Still more issues, however, remain unresolved; these include:

- Effects of G.992.2 power management, which may put the modems into a "sleep mode", where they would not be able to accept incoming calls, respond to OAM cells, or acknowledge ILMI "keep alive" messages. One simple solution would be to disable the G.992.2 power-down mode, but that would defeat one of the primary purposes of the recommendation for customer premise equipment!
- Can any ATM traffic class other than UBR be supported by G.992.2? If there is no splitter or minifilter, then whenever a phone/fax machine goes off hook[6], a G.992.2 modem may need to do a *fast retrain* to a lower upstream rate with lower transmit power[7], to prevent degradation of the voice quality (see Section 9.1.3). Even worse, if the ring trip or dial pulsing transients cause a loss of synchronization, a full retrain will be needed. Fast and full retrains, as presently defined, take 2 to 3 and 10 to 12 seconds, respectively, and they "take down" the PHY layer, thus breaking a traffic contract. Thus bandwidth for CBR and real-time VBR cannot be guaranteed, and only UBR traffic would be possible.

  NOTE: This makes a strong case for the use of minifilters, as described in Section 9.1.6.

- SNMP as an ADSL facility network management protocol? Note that G.997 and Annex L of T1.413 specify use of SNMP over a "clear" *embedded operations channel* for both regular and lite.
- New ATM UNI signaling (information elements?) for dual latency path selection and RR between the fast and interleave paths.
- Autoprovisioning of PVCs and SVCs to permit:
  - New PVCs to be created or modified
  - ADSL NT self-discovery of configuration parameters
  - ATM addresses of potential destinations (for SVCs)

2.6 ADSL NETWORK MANAGEMENT AND MANAGEMENT INFORMATION BUSSES

All three ADSL standards and recommendations (T1.413 and G.992.1 and.2) define *physical layer management* capabilities. These include a parameter

---

[6] The channel will also change (improve) when a phone goes back on hook, but it is unlikely that protocols will be developed in the near future to take advantage of increased channel capacity.
[7] This power cutback needs to be performed autonomously, because the higher layers cannot know when a phone goes off hook. Methods of detecting a change of impedance of the line have been proposed.

exchange at modem startup (discrete tones for T1.413, and G.994 for G.992); bidirectional indicator bits, which report receiver status every superframe (17 ms); and an embedded operations channel (eoc) for in-service testing and selected measurements. The indicator bits, eoc and an *ADSL overhead channel* (aoc), are contained within each superframe. *Performance monitoring* (PM) is also specified in these standards; it is mandatory for the ATU-C and optional for the ATU-R. The detailed PM aspects of ADSL in general and G.992.2 in particular will be covered in an appendix to a revision of T1.231 [ANSI, 1993a]. *Near-end PM* is defined as what the receiver observes and detects; *far-end PM* is what the (remote) far end detects and sends back via indicator bits. Both near- and far-end PM are mandatory at the ATU-C.

The ADSL Forum has standardized TR-006–ADSL Line MIB [ADSLF, 1998] for exchange of SNMP messages between an EMS (SNMP manager) and a DSLAM (SNMP agent), at the Q reference point. There is a modified version of that MIB specified in G.997 for use over the G.992 facility. The NM protocols for G.997 are SNMP over byte-oriented HDLC frames over a "clear eoc". No UDP or TCP/IP is required. However, the definitive ADSL MIB is likely to come from the IETF, which is the guardian of SNMP MIBs. A draft IETF MIB for ADSL is currently being reviewed.[8]

Other ADSL Forum NM specifications include:

- WT-022 (DMT line code-specific MIB)
- WT-023 (CAP line code-specific MIB)
- WT-025 (CMIP-based network management framework)

While WT-22 is essentially an extension of TR-006 ADSL line MIB, it is not clear either how or if the latter two ADSL Forum NM specifications will be used. The ATM Forum's ILMI (SNMP over AAL5) MIB will also be needed for ATM over ADSL, but the managed objects defined will not be specific to ADSL. A proposal to extend ILMI for autoprovisioning of PVCs will probably be accepted.

Summary. The ADSL facility is managed using ADSL PHY layer management (G.997 for G.9.992). The ATM aspects over ADSL will be managed by OAM cells (I.610) and the ILMI (SNMP). The EMS-to-DSLAM NM will be via either the ADSL Forum's line and DMT MIBs (SNMP) or, when completed, the IETF's ADSL MIB.

---

[8] This document defines a standard SNMP MIB for ADSL lines based on the ADSL Forum standard data model. The ADSL standard describes ATU-C and ATU-R as two sides of the ADSL line. This MIB covers both ATU-C and ATU-R agents' perspectives. Each instance defined in the MIB represents a single ADSL line. It should be noted that much of the content for the first version of this document came from work completed by the ADSL Forum's network management working group and documented in [ADSLF, 1998].

2.7 OBSERVATIONS

ADSL, both regular and lite, has the potential to provide very cost-effective high-speed Internet and remote access for residential and SOHO users. For this objective to be realized, new interfaces, equipment, and protocols will be needed. Standardized network management tools must be in place for configuration/ auto provisioning, fault detection, and performance monitoring. Further, we firmly believe that SVCs, in conjunction with both PPP and classical ATM protocol stacks, will be necessary to achieve scalable networks. QOS and point-to-multipoint virtual circuits would permit ADSL to be an enabling technology for video retrieval, video streaming, digital voice (VToA and VoIP), and multimedia conferencing. Let us hope that the ADSL and ATM Forums can work together to resolve many of the open issues identified here. Doing so will greatly increase ADSL's commercial success and viability.

3

THE DSL AS A MEDIUM FOR HIGH-SPEED DATA

Subscriber loops[1] which connect the customer premises to a central (or switching) office (CO), were developed and deployed for voice transmission, and have been well described by many authors. [Gresh, 1969], [Manhire, 1978], [Freeman, 1981], and [AT and T, 1982] are excellent references; they are old, but then the subscriber loop is very old, and not much has changed in twenty years! A recent description of those characteristics of the loops that are appropriate for DSL appears in [Rezvani and Khalaj, 1998].

NOTE: With the advent of fiber to the neighborhood (FTTN: see Section 1.2), subscriber loops will also be used to connect customer premises to an optical network unit (ONU) using VDSL. When describing the use of loops for generic DSL, I will refer only to CO and will differentiate between CO and ONU only when discussing VDSL specifically.

3.1 MAKE-UP OF A LOOP

Each subscriber loop consists of a pair of insulated copper wires of gauges ranging from 26 AWG to 19 AWG (approximately 0.4 to 0.91 mm). The insulating dielectric is mostly polyethylene, but some paper-insulated pairs still exist. A typical loop plant, as shown in Figure 3.1, consists of a multipair feeder cable emanating from the CO; this may contain up to 50 *binder groups*, each of which may contain 10, 25, or 50 pairs. At a *feeder distribution interface* (FDI) a feeder cable is then divided into several smaller (up to 50 pairs) distribution cables; these are then finally broken out into many individual drop-wire pairs to customer premises.

Within the cables the two wires of each pair are twisted around each other to form an unshielded (and unsheathed) twisted pair (UTP). ANSI is presently

---

[1] See Section 3.1.3 for why it is called a loop.

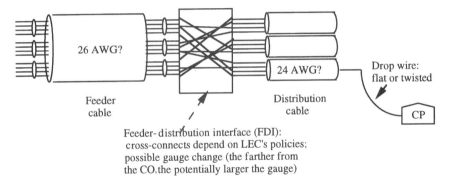

26 AWG?

24 AWG?

Drop wire:
flat or twisted

Feeder
cable

Distribution
cable

CP

Feeder-distribution interface (FDI):
cross-connects depend on LEC's policies;
possible gauge change (the farther from
the CO.the potentially larger the gauge)

Figure 3.1 Typical loop plant: feeder and distribution cables to customer premises.

defining the properties (twist length or *pitch*, balance, dielectric loss, etc.) of several categories of UTP: Category 3 and Category 5 in particular. Most of the installed plant is Cat-3 or lower (i.e., worse in some or all properties), but there is a small amount of Cat-5 installed, particularly from ONUs to new customer premises. The pitch of Cat-3 can vary from about 1.5 to 3 ft, and the twist is hardly discernible to the untrained eye when the outer sheath of a cable is removed. For purposes of maintaining balance, however the most important parameter is the ratio of the signal wavelength to the pitch; even at 15 MHz, which is about the highest frequency presently contemplated for use on UTPs, the wavelength/pitch ratio is about 20:1. The pitch for Cat-5 is only a few inches and is precisely varied from pair to pair within a cable; the crosstalk balance may be as much as 20 dB better than for Cat-3.

[Rezvani and Khalaj, 1998] report that in the United States most multipair cables are constructed in an attempt to make all pairs "equal"; that is, the position of any pair within the cable changes, and no two pairs stay close together for any great distance; this is intended to average the crosstalk between different pairs and to reduce the difference between the worst- and best-case interferers (see Section 3.6). I have, however, also heard the opposite opinion: that pairs tend to maintain their position in a cross section. There may well be both types of cable out there, making the task of modeling (see Section 3.6) even more difficult. In other countries (e.g., Japan and Germany) two pairs are first twisted as quads, which are then combined in a larger cable. The crosstalk between pairs in the same quad is much higher than average, and that between pairs in different quads is lower than average.

3.1.1 Length of the Loop

Telephone plants throughout the world vary widely in the distribution of their customers (i.e., in percentage of customers covered as a function of distance from the CO). During the development of T1.413 it was generally "agreed" that the so-called "extended carrier serving area" with a nominal 18-kft radius would include about 80% of all customers; this was consistent with Bellcore's 1973 loop survey [AT&T, 1982], which showed 85% within 18 kft. It was probably tacitly assumed, moreover, that the remaining 20% were typically in rural areas with a lower demand for data services. As a counterexample, however, one central office in San Jose, California (a modern city in Silicon Valley with highly sophisticated data-hungry residents) has approximately 64% of its customers *more* than 18 kft away.

3.1.2   Balance

All signals on the subscriber loop are carried in the differential mode[2], in which the current in one wire is balanced by an equal and opposite current in the other. Every effort is made—in both the manufacture of the cable and the design of the terminal equipment—to minimize the common-mode[3] component. Transmitters should be able to achieve a differential mode/common mode ratio of at least 55 dB across the used band, but because of imbalance of the two wires of any pair to "ground" (represented mainly by the other pairs), there is some differential mode-to-common mode conversion in the cable. For Cat-3, the most common type of UTP used in the United States, the output ratio is about 50 dB below 100 kHz and falls to about 35 dB at 10 MHz.[4]

3.1.3   Wire Gauge and Gauge Changes

The primary parameter that controls the ability of CO equipment to perform signaling and diagnostic maintenance is the dc resistance of the loop measured between the two wires at the CO, with the wires shorted at the customer premises. In the United States, according to the *revised resistance design* (RRD) rules, the loop resistance is limited to $1500\,\Omega$.[5] Therefore, the ideal arrangement would be to adjust the gauge of the wires according to the length of the loop: the longer the loop, the larger the gauge.

Such an ideal cannot be achieved in practice, however, because, as shown in Figure 3.1, different pairs (all necessarily of the same gauge) in a large feeder cable emerging from a CO might eventually go to premises at widely varying distances. Therefore, a common practice is to start out from the CO with feeder cables containing many fine-gauge pairs, and increase the gauge at an FDI as the distance from the CO increases. At least one gauge change, therefore, may occur within the feeder/distribution cables and must be considered in any mathematical analysis (see Section 3.5).

---

[2] Originally, differential-mode current was called *metallic circuit* current to distinguish it from the common-mode current, which used a ground (i.e., nonmetallic) return.

[3] Originally called *longitudinal* mode.

[4] More on this in Sections 3.5 and 9.3.2.

[5] Eighteen kilofeet of 26 AWG has a resistance of $1500\,\Omega$.

3.1.4 Bridge Taps

Bridge taps are open-circuited lengths of UTP that are connected across the pair under consideration. They can be the result of many different installation, maintenance, and house wiring practices:

- *Party lines.* In the early days of telephony it was common for several customers to share the same pair. Then when more cables were installed and privacy became more affordable, the drops to the other premises were just disconnected, leaving the unterminated pairs (open-circuit stubs) still connected to the used loop. A simple configuration is shown in Figure 3.2(*a*); a more complicated and less common one, with a bridge tap on a bridge tap, is shown in Figure 3.2(*b*).
- *Extension of the distribution cable beyond the drop to the customer premises.* According to [AT&T, 1982], "the cable pair serving the customer *usually* [my emphasis] extends past the customer to the point at which the particular cable run ends." These are sometimes called *tapped-in drops.*
- *Repairs.* If a pair breaks somewhere inside a cable, the repairer may simply splice in another pair without disconnecting the broken sections. It can be seen from Figure 3.3 that this leaves two bridge taps connected to the loop in use.

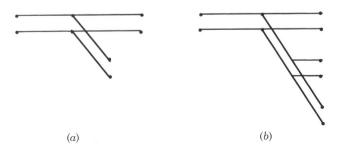

(*a*)   (*b*)

Figure 3.2 Bridge taps: (*a*) simple; (*b*) bridged.

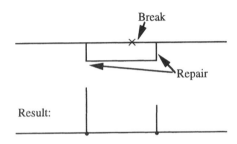

Figure 3.3 Two bridge taps caused by a repair.

- *Extra capacity.* To allow for future service to any one of several potential customer premises, it is common practice to splice one pair in a feeder cable to one pair in each of several distribution cables. The unused pair(s) then form bridge tap(s).

- *Multiple telephone outlets within customer premises.* The most common in-house wiring configuration is a tree with its base at the service entrance. All branches that are either unterminated or teminated in on-hook telephones constitute short bridge taps that may be significant at VDSL frequencies. This is discussed in more detail in Chapter 10.

3.1.5 Loading Coils

A loop is often thought of as having a bandwidth of only 4 kHz, but that limitation is imposed by multiplexing equipment on the network side of the CO; it is not inherent in the loop itself. Because the switched telephone network (STN), which interconnects COs, originally used frequency-division multi-plexing based on multiple 4-kHz bands,[6] the signals within the STN must be bandlimited to something less than 4 kHz. Therefore, if the subscriber loop is to be used only for access to the STN, there is no need for a bandwidth greater than 4 kHz.

At low frequencies UTP acts like a distributed RC circuit, and its response droops across the 4-kHz voice band (by as much as 12 dB on long loops). That droop reduced the capacity of early telegraphy systems and degraded the voice quality, so Heaviside[7] proposed that lumped inductors be added in series at regular intervals along the loop. A common configuration in the United States is 88-mH coils inserted every 6000 ft; a 26-AWG loop so loaded would be designated 26H88. These *loading coils* ideally convert a droopy RC network into a maximally-flat low-pass filter with a cutoff around 3.0 kHz. In the process of improving the voice-band response, however, loading coils greatly degrade the response beyond 4 kHz, so they must be removed (or perhaps just shorted out) to allow any wider-band service to operate on the loop.[8] Removing the coils may be a significant part of the cost of providing DSL service.

3.1.6 The Drop Wire

When a pair finally emerges from a distribution cable, it is connected to the customer premises by a *drop wire.* The term refers to the "drop" from a pole, which often occurs even if the distribution cable is underground. Drop wires may be copper, steel, or a mixture. They may be flat or twisted, and their balance is usually much worse than that of the UTP part of the loop. This may result in

---

[6] And has now largely converted to digital systems based on 8-kHz sampling.
[7] See [Riezenman, 1984] for an interesting story about the invention of loading coils.
[8] It is ironic that loading coils, which were originally added to increase the capacity of loops, must now be removed to increase it further!

pickup of radio-frequency (RF) noise (see Section 3.7.1). The characteristic impedance of drop wires is typically higher than that of UTP, and the result of the impedance mismatch on the attenuation may be significant on the shorter loops and at the higher frequencies used for VDSL. They were ignored in the definition of the test loops for ADSL, but are considered for VDSL.

3.2   LADDER MODEL OF AN UNSHIELDED TWISTED PAIR[9]

NOTE: This section contains *much* more detail than most readers will want, but I included it all because such a level of detail will be needed for the study of crosstalk cancellation, and because I would like this to be a comprehensive description of UTP used for xDSL.

A UTP comprises distributed inductance and resistance in series, and distributed capacitance and conductance in shunt. All four *primary* parameters are cited per unit length (kft in the United States; km elsewhere). A homogeneous section of unspecified "unit" length and a cross section of a pair are shown in Figures 3.4(*a*) and (*b*).

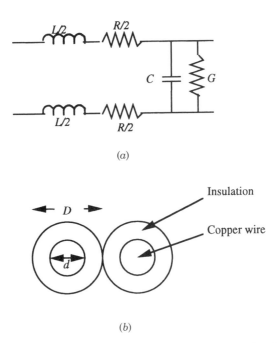

(*a*)

(*b*)

Figure 3.4   (*a*) Lumped model of unit length; (*b*) cross section of UTP.

[9] If the model is used only for the differential mode, it is equally valid for both UTP and flat pairs; nevertheless, for simplicity we refer only to UTP from here on.

The capacitance per unit length is given by

$$C_{per} = \frac{\pi k \varepsilon_0}{\text{arccosh}(D/d)} \qquad \text{F/m} \qquad (3.1)$$

where k is the dielectric constant of the medium and ε_0, the permittivity of free space, 8.85×10^{-12}. This formula assumes that the insulating medium is homogeneous, but in practice it is not; there are two sheaths of insulator as shown, and beyond that, an unknown mixture of air and the insulation of other pairs. The dielectic constant of polyethylene $= 2.26$, but the effective k value, which probably varies slightly with cable makeup, appears to be ≈ 2.05.

The inductance per unit length at high frequencies — when the current is carried mostly on the surface of the wires—is given by

$$L_{per} = \frac{\mu_0}{\pi} \text{arccosh}\,(D/d) = \frac{\mu_0}{\pi} \ln\left[\frac{D}{d} + \sqrt{\left(\frac{D}{d}\right)^2 - 1}\right] \qquad \text{H/m} \qquad (3.2)$$

where μ_0, the permeability of free space $= 4\pi \times 10^{-7}$. Hence

$$Z_0 = \sqrt{\frac{L_{per}}{C_{per}}} = \sqrt{\frac{\mu_0}{k\varepsilon_0}} \frac{\text{arccosh}(D/d)}{\pi} \qquad \Omega \qquad (3.3)$$

For polyethylene insulated cable (PIC) in our peculiar North American units,

$$C_{perkft} = \frac{17.6}{\text{arccosh}(D/d)} \qquad \text{nF/kft} \qquad (3.4)$$

$$L_{perkft} = 0.122\,\text{arccosh}(D/d) \qquad \text{mH/kft} \qquad (3.5)$$

and for 26 AWG, $D/d \approx 1.7$.

Two other formulas for C_{per} and L_{per} have been used in the literature, with the arccosh replaced by $\ln(2D/d-1)$ or $\ln(2D/d)$. Both are valid approximations if $D/d \gg 1$, but that is not true for UTP. The former would be exact if the current were uniformly distributed: either throughout the wire at low frequencies or around the surface at high frequencies. In practice, however, the effect of the EM field created is to concentrate the currents closer to the other wire, and the arccosh forms take account of this; the values of capacitance and inductance given by (3.1) and (3.2) are about 20% higher than those given by $\ln(2D/d-1)$. The $\ln(2D/d)$ form is used mainly because of its simplicity; the resulting values are about 7% higher than the exact ones.

Inductance at dc. The L_{per} value given by (3.2) is often called *external* because it results from flux linkages outside the wire. At low frequencies, when the current flows through the full cross section of the wire, there is also an *internal* inductance,

$$L_{int} = \frac{\mu_0}{4\pi} \qquad (3.6)$$

and the L_{per} value at low frequencies is the sum of the internal and external.[10]

Each homogeneous section of UTP can also be characterized by a pair of *secondary* parameters: Z_0, the characteristic impedance, and γ the propagation constant. γ would exactly define the propagation of a single section of homogeneous UTP if it were terminated (at both ends) in its characteristic impedance. Z_0 is, however, complex and frequency-dependent, and such *matched* terminations can only be approximated. Nevertheless, even with purely resistive terminations, γ is a very accurate ($\pm 0.2\,dB$) predictor of the attenuation of a single section at frequencies above about 20 kHz; in fact, the sum of the gammas for a tandem connection of mismatched in-line sections is also accurate. Most end-to-end transmission paths, however, include lumped elements (transformers, etc.) and bridge taps. The propagation constants of each section are not by themselves sufficient for an accurate analysis of such a loop; an analysis method such as that described in Section 3.5 is necessary.

Z_0 and γ are therefore used mainly as intermediate parameters that define a set of *tertiary* parameters, such as the elements of the chain matrix of the lumped-element section of Figure 3.4; these in turn allow exact analysis of any number of mismatched sections with any terminations. Explicit knowledge of $\gamma(= \alpha + j\beta)$ is, however, useful in the early stages of design of DSL systems for two reasons:

1. The rate of change of β, the imaginary part of γ, defines the propagation delay of each section at any frequency:

$$\tau = \frac{d\beta}{d\omega} \qquad s/\text{unit length} \qquad (3.7)$$

   A plot of τ versus frequency for various gauges is shown in Figure 3.5. It can be seen that the average delay is about 1.5 µs/kft; that is, the propagation velocity on UTP is about 65% of that in free space. These delays can be summed to give the approximate propagation delay of the entire loop (see Sections 4.2 and 10.5 for how this is needed).

2. An estimate of the attenuation of a section will be useful in Section 3.5.2 when considering the effects of bridge taps. Above about 300 kHz the attenuation per unit length can be approximated by

$$dB = 8.686\alpha(f) \approx \alpha_1 \sqrt{f} \qquad (3.8)$$

---

[10] L_{int} and L_{ext} do not quite add up to L_0, so there is some other effect here that I do not understand.

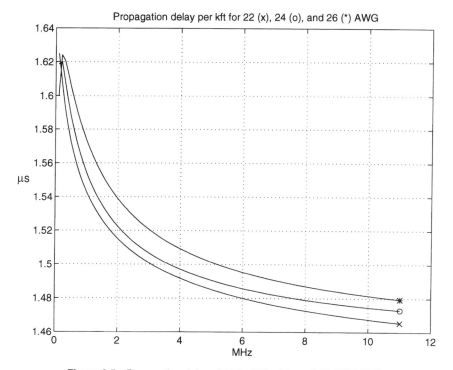

Figure 3.5 Propagation delay of 1 kft of 22-, 24-, and 26-AWG UTP.

Values of α_1 (normalized to 1 MHz) for several American, European, and Japanese cables are given in Tables 3.1 and 3.2.

3.2.1 Is a UTP a Minimum-Phase Network?

The question of whether a UTP is a minimum-phase network becomes important when the design of equalizers is considered. It appears from the equivalent circuit of Figure 3.4 that there are no transmission zeros in the right-half p-plane (in fact, they are all at infinite frequency). This would certainly suggest that a UTP is a minimum-phase network (see pp. 303–309 of [Bode, 1945]), but that is a shaky argument because the network is distributed. In fact, in the strictest sense, a UTP cannot be minimum phase because of the "excess" phase (i.e., over the phase associated with the attenuation) caused by the propagation delay. John Cook of BT has suggested[11] that if this frequency-invariant delay (linear phase change) is subtracted,[12] the remainder is *quasi minimum phase*. This might be an interesting subject for study.

---

[11] Private correspondence.
[12] Such a delay is unimportant in the design of equalizers.

3.3 DISTRIBUTED *RLGC* PARAMETERS

NOTE: I am very indebted to Joseph Musson of Marconi and John Cook of British Telecom for enlightening discussions on this subject. Most of the good stuff in this section comes from them.

UTPs can be characterized by measuring some *tertiary* parameters—either the input open- and short-circuit impedances (Z_{oc} and Z_{sc}) or the scattering parameters — of a unit length at various frequencies. From these Z_0 and γ can be calculated, and thence R, L, G, and C. Both sets of calculations are described in detail in [Pollakowski, 1995]. Musson has pointed out,[13] however, that these calculations — particularly those that use only Z_{oc} and Z_{sc} — are very sensitive to perturbations in the measurements due either to noise (generally avoidable by careful measurement techniques) or to nonuniformity in the cable (generally unavoidable); as a result, extreme swings of L and C may be indicated. Some smoothing from frequency to frequency is therefore essential, but there is as yet no universal agreement on how this should be done.

Tables of the calculated values of R and L up to 5 MHz were published in [Bellcore, 1983], but it is not clear how they were smoothed. More comprehensive tables were published in [Valenti, 1997]; for these, Z_0 and γ were smoothed as described in [ASTM, 1994].

For all PIC cables G is negligible over the xDSL frequency range (<15 MHz), and, for U.S. cables at least, C is essentially constant with frequency; L and R are the only parameters of interest. [Cook, 1996] proposed a two-coefficient function[14] for R:

$$R(f) = R(0)\left[1 + \left(\frac{f}{f_r}\right)^2\right]^{0.25} \qquad \text{k}\,\Omega \text{ per kft (U.S.) or per km} \qquad (3.9)$$

and a four-coefficient function for L:

$$L(f) = \frac{L(0) + L(\infty)x_b}{1 + x_b} \qquad \text{mH per kft or per km} \qquad (3.10)$$

where

$$x_b = \left(\frac{f}{f_m}\right)^b \qquad (3.11)$$

---

[13] [Musson, 1998] and private correspondence.
[14] Cook's original formula for R, which is used in [Cioffi, 1998], allows for a steel reinforcement (used in some drop wires for added tensile strength) that may also conduct. The resistance of the steel part is, however, specified as infinite, so there seems to be no need for the more complicated formula.

NOTE: Equation (3.9) is different in form (but not in substance) from Cook's original formula; it makes scaling between kft and km much simpler. Unfortunately, the original formula has been perpetuated in G.996.

Equations (3.9) and (3.10) can be fitted in many different ways — minimizing the error in the absolute values of R and L or in the relative values [i.e., error $=$ $(R - R_{\text{fit}})/R$], minimizing the mse or maximum error, spacing the frequencies on a linear or log scale—and all will give slightly different sets of $R(0)$, and so on. Furthermore, one can never be sure that one has found the global minimum of the error. Nevertheless, the primary parameters of PIC-insulated U.S. cables (see [AT&T, 1983]) and CCP-insulated Japanese cables[15] over the xDSL frequency range 30 kHz to 10 MHz can be fitted[16] closely enough for practical purposes. Values of $R(0), f_r, L(0), L(\infty), f_m$, and b are shown in Table 3.1.

Two slightly different sets of parameters have been published for the British Telecom cables used in the United Kingdom. The first — for the ADSL frequency range — were based on primary parameters reported in Annex H of T1.413. The second — for the VDSL frequency range — were reported in [Cioffi, 1998]. Both are shown in Table 3.2, but there is a mismatch of more than 10% between them in the overlapping region around 1 MHz.

3.3.1 *R* and *L*, and *G* and *C* as Hilbert-Transform Pairs

[Musson, 1998] pointed out that because R and $j\omega L$ are the real and imaginary parts of a physical impedance, they must be related by the Hilbert transform. He showed that independent "best" fits of R and L are not so related, and for some UTPs the values of L calculated as a Hilbert transform of an R that is fitted according to (3.9) differ significantly from the values of L fitted according to (3.10)! The problem seems to lie in the inaccuracy of (3.9) beyond the band of interest and in the effect of this on the integration involved in the Hilbert transform.

To solve this problem, Musson studied the physics of a copper pair as discussed in [Schelkunoff, 1934], [Kaden, 1959], and [Lenahan, 1977], and defined the combined impedance of the series branch in Figure 3.4 as a single function of the variable $s = j\omega$:

$$Z_{\text{ser}} = sL + R \tag{3.12}$$

$$= sL(\infty) + R(0)\left[0.25 + 0.75\sqrt{1 + \frac{as(s + b)}{(s + c)}}\right] \tag{3.13}$$

and similarly the admittance of the shunt branch

$$Y_{\text{sh}} = sC(1\,\text{MHz})\left(\frac{s}{2\pi j}\right)^{-2s/\pi} \tag{3.14}$$

---

[15] NEC, private correspondence.

[16] Minimizing the maximum relative error with frequencies spaced logarithmically.

TABLE 3.1 Coefficients for the Models of R and L for U.S. and Japanese Cables

Cable	$R(0)$	f_r	Max. % Error in R	$L(0)$	$L(\infty)$	f_m	b	Max.% Error in L
U.S. (per kft)								
22 AWG PIC at 70°	0.0342	0.0976	2.66	0.196	0.129	0.328	0.575	0.95
24 AWG PIC at 70°	0.0538	0.151	2.55	0.186	0.134	0.665	0.831	0.81
26 AWG PIC at 70°	0.0833	0.226	2.37	0.187	0.135	0.860	0.841	0.47
Japanese (per km)								
0.65 mm CCP	0.106	0.0725	2.48	0.857	0.532	0.129	0.892	0.98
0.5 mm CCP	0.179	0.121	2.25	0.968	0.527	0.0975	0.736	1.27
0.4 mm CCP	0.272	0.182	2.03	1.089	0.550	0.0748	0.666	1.57

Notes:
1. The plots of fitting error with frequency for all the cables are both smooth and similar. It is clear that the errors are not random; they are due either to systematic errors in the original measurements, or, more likely, small inadequacies of the model (particularly the one for resistance).
2. The coefficients for the U.S. cables are slightly different from those given in G.996, because the frequency range has been extended to 10 MHz to cover all xDSL.

TABLE 3.2   Coefficients for the Models of *R* and *L* for BT Cables

Cable	$R(0)$	f_r	Max. % Error in R	$L(0)$	$L(\infty)$	f_m	b	Max. % Error in L
T1.413 (per km)								
0.63 mm	0.113	0.0797	0.0274	0.702	0.473	0.270	1.053	0.066
0.5 mm	0.179	0.135	0.0224	0.673	0.538	0.638	1.273	0.091
0.4 mm	0.280	0.252	0.048	0.587	0.436	0.678	1.456	0.068
VDSL (per km)								
TP2 (0.5 mm)	0.174	0.132	?	0.617	0.479	0.554	1.153	?
TP1 (0.4 mm)	0.286	0.213	?	0.675	0.489	0.806	0.929	?

Notes:
1. The fit for the T1.413 cables is extremely good (error < 0.1%!), but it should be noted that the *R* and *L* values listed in Annex H of T1.413 are not "raw"; they had already been smoothed according to some formulas very much like (3.8) and (3.9). Their fit therefore is no validation of the models.
2. The coefficients for the ADSL cables are slightly different from those given in G.996 because the fitting has been improved since the ITU numbers were finalized.

NOTE: Equations (3.13) and (3.14) are slightly different in form (but not in substance) from Musson's formulas; they use *s* as the single variable throughout.

Musson's argument is that such mutually transformable pairs allow better smoothing of "noisy" values corrupted by measurement errors. This would seem to be more important for cables or frequencies for which *G* is significant and/or *C* varies with frequency. For many UTPs below about 10 MHz, however, $G \approx 0$ and *C* is constant; under those conditions the calculation of *R* and *L* is much less sensitive to measurement perturbations, and (3.9) and (3.10) may be adequate.

3.3.2   A Recommendation

As DSL systems are deployed more and more, many different cable types will be encountered, and their *RLGC* parameters will be important to designers. Publishing large tables of "raw" parameters would clearly be impractical, but smoothing (filtering) according to some favorite, nonstandard algorithm would be misleading. Therefore, some standards body should take it as a project to:

1. Establish the criteria for a fit
2. Examine all candidate smoothing algorithms[17]
3. Choose one
4. Define it precisely

---

[17] [ETSI, 1998], which, for the most part, is the European equivalent of the ANSI VDSL requirements document [Cioffi, 1998], also defines a model proposed by KPN, which should be considered a candidate.

5. Liaise with other standards bodies to ensure that all referenced cables are defined by their primary parameters, which have been derived in this way

3.4 TRANSFORMER COUPLING AND dc BLOCKING

All subscriber loops must be transformer-coupled in order to protect terminal equipment from large common-mode voltages that might be induced on the loop. All digital equipment must also be capacitively coupled to protect it from the dc current (up to 100 mA on some short loops) used for signaling. The series capacitor and shunt inductor of the transformer comprise a minimum high-pass filter at each end, so that the end-to-end transfer function must have at least four zeros at zero frequency; this will be significant when considering equalizers in Chapter 8. A typical end-to-end loop, with one bridge tap, one gauge change, and one drop wire, is shown in Figure 3.6.

3.5 CHAIN MATRIX CHARACTERIZATION

A subscriber loop comprising in-line sections, bridge taps, a drop wire, and lumped elements can be analyzed most easily via its chain matrix. The chain matrices of the separate components are defined as follows.

3.5.1 In-line Sections

A section of homogeneous, symmetrical UTP of length l can be characterized as

$$M_i = \begin{vmatrix} A_i & B_i \\ C_i & D_i \end{vmatrix} = \begin{vmatrix} \cosh(\gamma_i l_i) & Z_0 \sinh(\gamma_i l_i) \\ Y_0 \sinh(\gamma_i l_i) & \cosh(\gamma_i l_i) \end{vmatrix} \qquad (3.15)$$

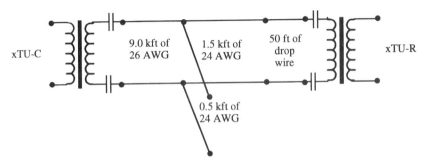

Figure 3.6 Typical end-to-end connection.

where

$$Z_0 = \sqrt{\frac{R + j\omega L}{G + j\omega C}} \qquad (3.16)$$

$$\gamma = \sqrt{(R + j\omega L)(G + j\omega C)} \qquad (3.17)$$

NOTE: Because all components of the loop are passive, the determinants, $A_i D_i - B_i C_i$, of all the factor matrices, M_l, and of their product, M, are unity.

3.5.2 Bridge Taps

Bridge taps present their open-circuit admittance, Y_{oc}, in shunt between two in-line sections. The chain matrix for a bridge tap is therefore

$$M_{bt} = \begin{vmatrix} 1 & 0 \\ Y_{oc} & 1 \end{vmatrix} \qquad (3.18)$$

For the simple bridge tap of Figure 3.2(a),

$$M_{bt} = \begin{vmatrix} 1 & 0 \\ Y_0 \tanh(\gamma l) & l \end{vmatrix} \qquad (3.19)$$

For the "bridged" bridge tap of Figure 3.2(b), M_{bbt} is also given by (3.18) with

$$Y_{ocbbt} = \frac{Y_{01} \tanh(\gamma_1 l_1) + Y_{02} \tanh(\gamma_2 l_2) + Y_{03} \tanh(\gamma_3 l_3)}{1 + Z_{01} \tanh(\gamma_1 l_1)[Y_{02} \tanh(\gamma_2 l_2) + Y_{03} \tanh(\gamma_3 l_3)]} \qquad (3.20)$$

3.5.3 High-Pass Filters

The high-pass filters at each end comprise a series and a shunt element, so that

$$M_{hp1} = \begin{vmatrix} 1 & 1/j\omega C_{hp} \\ 1/j\omega L_{hp} & 1 - 1/\omega^2 L_{hp} C_{hp} \end{vmatrix} \qquad (3.21)$$

$$M_{hp2} = \begin{vmatrix} 1 - 1/\omega^2 L_{hp} C_{hp} & 1/j\omega C_{hp} \\ 1/j\omega L_{hp} & 1 \end{vmatrix} \qquad (3.22)$$

3.5.4   The End-to-End Loop

For the loop shown in Figure 3.6,

$$M = \begin{vmatrix} A & B \\ C & D \end{vmatrix} \tag{3.23}$$
$$= M_{hp1} M_1 M_{bt} M_2 M_{dw} M_{hp}$$

The transfer function of the end-to-end loop is

$$H = \frac{2}{A + BG_2 + CR_1 + D} \tag{3.24}$$

and the input impedances at the two ends of the loop (needed for the design of the echo canceler) are

$$Z_1 = \frac{A + BR_2}{C + DR_2} \tag{3.25}$$
$$Z_2 = \frac{D + BR_1}{C + AR_1} \tag{3.26}$$

3.5.5   MATLAB Program for Chain Matrix-Based Analysis

A MATLAB program to perform the above analysis is given in Appendix A. The results of the analysis of the loop of Figure 3.6 without the high-pass filters,[18] but both with and without the 500-ft bridge tap, are shown in Figure 3.7. This also shows the attenuation of the loop without the bridge tap, as calculated from the real part of γ; above about 20 kHz the match is very good.

NOTE: The even simpler formula of attenuation, $\alpha\sqrt{f}$, is accurate only above about 300 kHz when the skin effect upon resistance becomes dominant.

3.5.6   Frequency and Depth of the Notch Caused by a Simple Bridge Tap

An open-circuited section of cable has minima of its input impedance (producing notches in the end-to-end transfer function) at frequencies for which the length is an odd number of quarter-wavelengths. The propagation velocity of U.S. cables over the ADSL frequency range[19] is approximately 0.63 kft/μs ($\pm$ 5%).

---

[18] The filters will vary from one implementation to another.
[19] The velocity may have to be adjusted slightly for other cables or for the VDSL frequency range.

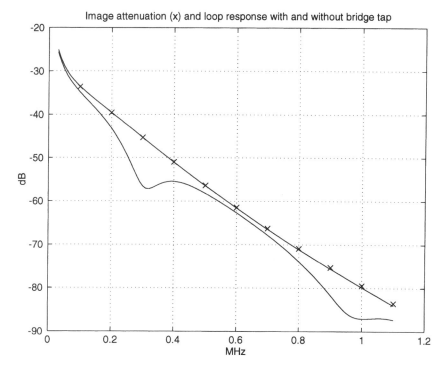

Figure 3.7 Frequency response of connection (without high-pass filters) shown in Figure 3.6.

Therefore, the first notch frequency is

$$f_{\text{notch1}} \approx \frac{0.63}{4l}$$

$$\approx 0.315 \, \text{MHz} \qquad \text{for a 500-ft bridge tap} \tag{3.27}$$

The signal that is propagated down the bridge tap and reflected back is attenuated, as given by (3.8), by

$$\text{att} \approx \alpha \cdot 2l\sqrt{f} \qquad \text{dB} \tag{3.28}$$

For the 24-AWG bridge tap of this example, $\alpha \approx 6.2 \, \text{dB/kft}/\sqrt{\text{MHz}}$, so $\text{att}(f_{\text{notch1}}) \approx 3.48 \, \text{dB}$.

The depth of the notch is then given by

$$\text{dB}_{\text{notch1}} = 20 \log(1 - 10^{-\text{att}/20})$$

$$\approx 9.6 \, \text{dB for a 500-ft 24-AWG bridge tap} \tag{3.29}$$

As can be seen from Figure 3.7, the notch is at approximately 0.31 MHz, and the attenuations there with and without the bridge tap are approximately 56 and

46 dB, so the estimates of notch frequency and depth are very good. The secondary notch, for which the length is three quarter-wavelenghts, is, as would be expected, at approximately 0.9 MHz, but the estimation of the notch depth there is much more complicated.

Multiple Bridge Taps. Loop engineering practices often specify some maximum total length of all bridge taps on a loop; a figure of 2000 ft is often cited. For xDSL use this is a completely misleading parameter; many short bridge taps are *much* more harmful than one long one.

3.5.7   Calculated Versus Measured Responses: A Cautionary Tale

Most engineers developing xDSL systems have probably gone through the following sequence of thoughts and experiences:

1. xDSL equipment should be checked with real loops.
2. Very few development labs are 12 kft long!
3. For convenience the cable can be left on the drum.
4. However, when the attenuation is measured with the cable wound in many loops on a drum, crosstalk from the early loops of cable, in which the signal level is high, into the later loops, in which the signal is attenuated and the level low, overwhelms the real transmitted signal.
5. Therefore, the cable must be spread out, but looped back on itself so that the input and output can be connected to the same measuring instrument.

Laying the cable out in one big loop is certainly much better than leaving it wound in many layers on a drum, but there are still potential problems with such an arrangement. Figure 3.8, which may be distressingly familiar to some DSL engineers, shows the calculated effects of just a 1-pF parasitic capacitor connected from input to output of a 4-kft 24-AWG loop. Clearly, from about 5 MHz onward the lossy loop is bypassed by the capacitor, and measured results would be useless.

3.6   CROSSTALK

Crosstalk between pairs in a multipair cable is the dominant impairment in any DSL system. The cause of this crosstalk is capacitive and inductive coupling (or, more precisely, imbalance in the couplings) between the wires. In Sections 3.6.1 to 3.6.4 we consider the traditional statistical models of crosstalk. Precise knowledge of individual pair-to-pair crosstalk transfer functions will, however, be needed in any effort toward crosstalk cancellation (see Section 11.5), and a proposed model and method of analysis are discussed in Section 3.6.5.

If one pair is considered as the interferer, the voltages and currents induced in the other pairs travel in both directions; those that continue in the same direction

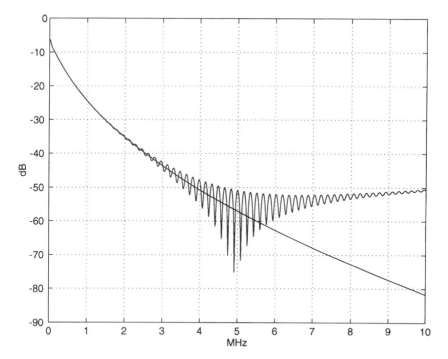

Figure 3.8   Response of 4 kft of 24 AWG with 1 pF of parasitic capacitance connected between input and output.

as the interfering signal add up to form far-end crosstalk (FEXT); those that come back toward the source of the interferer add up to form near-end crosstalk (NEXT). This is shown conceptually in Figure 3.9, where the thickness of the lines showing the crosstalk is a crude indication of the relative levels of the signals involved. If both NEXT and FEXT can occur in a DSL system, NEXT will in general be much more severe.[20] NEXT increases with frequency, and at VDSL frequencies (up to 15 MHz) it would be intolerable; therefore, VDSL systems are designed to avoid it altogether. The examples discussed in the next sections are of NEXT at ADSL frequencies (up to 1.1 MHz) and FEXT at VDSL frequencies (up to 15 MHz).

Until now worst-case values of the multiple-pair-to-pair crosstalk transfer functions have been the main interest; these have been used by DSL modem designers and service providers to predict (indeed, to guarantee) data rates and coverage, and by the standards bodies to define tests for DSL modems. Now, with the increasing ability of transmission protocols to use whatever data rate is available (see Chapter 2), statistical average values will also be interesting. Statistical models for both 1% worst-case and average values are discussed in Section 3.6.3. The models for the averages are only provisional, but they should be accurate enough for systems planning.

---

[20] But see the discussion of "Amplified FEXT" in Section 3.6.2.

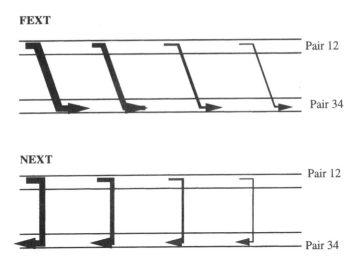

Figure 3.9   Inter pair coupling causing FEXT and NEXT.

3.6.1   NEXT

As shown in Figure 3.9, NEXT from pair 12 to pair 34[21] is the sum of an infinite number of small signals that are propagated some distance down loop 12, coupled across, and propagated back up loop 34. That is, somewhat imprecisely,

$$H_{\text{NEXT}}(l,f) = \int_0^l H_{f12}(\lambda,f)H_{\text{XT}}(\lambda,f)H_{b34}(\lambda,f)d\lambda \qquad (3.30)$$

where $H_{f12}(\lambda,f)$ is the forward transfer function of a length λ of pair 12, $H_{\text{XT}}(\lambda,f)$ is the cross-coupling function at a distance λ from the input, and $H_{b34}(\lambda,f)$ is the backward transfer function of pair 34.

NOTE: It is assumed for the moment that there is no interaction between the transfer and cross-coupling functions, but this simplifying assumption is removed in the modeling and analysis described in Section 3.6.5.

Equation (3.30) can be simplified if pairs 12 and 34 have no bridge taps (or have the same configuration of bridge taps); then their transmission characteristics are the same, and

$$H_{f12}(\lambda,f) = H_{b34}(\lambda,f) = H(\lambda,f) \approx e^{-\gamma\lambda} \qquad (3.31)$$

---

[21] This is a clumsy notation, but it will be useful when we need to differentiate between the wires of each pair.

Then

$$H_{\text{NEXT}}(l,f) = \int_0^l H(2\lambda,f)H_{\text{XT}}(\lambda,f)d\lambda \qquad (3.32)$$

It can be appreciated that because H has a frequency-dependent phase shift, the multiple contributions to H_{NEXT} at any frequency may add or subtract, and depending on the sign and magnitude of H_{XT} (l, f), may cause peaks or deep notches. Plots of $|H_{\text{NEXT}}(f)|^2$ for many pair-to-pair measurements shown in [Pollakowski, 1995] and simulations described in Section 3.6.4 all confirm this rather simplistic explanation.

If, however, the contributions are considered to add on a power basis, and H_{XT} is written as the dimensionless product of a cross-coupling susceptance and some unspecified load impedance, the crosstalk power transfer function can be written as

$$|H_{\text{NEXT}}(f,l)|^2 \approx \int_0^l |H(2\lambda,f)|^2(\omega\,CR)^2d\lambda \qquad (3.33)$$

Hence, if the attenuation is considered to be approximately proportional to $\sqrt{f}$,

$$|H_{\text{NEXT}}(f,l)|^2 \approx \int_0^l (2\pi CR)^2f^2e^{-4\alpha\sqrt{fl}}df = \frac{(2\pi CR)^2}{4\alpha}f^{1.5}(1 - e^{-4\alpha\sqrt{fl}})$$

$$(3.34)$$

The last term in parentheses tends to unity for "large"[22] values of l; $|H_{\text{NEXT}}(f,l)|^2$ therefore approaches an asymptotic value and becomes independent of the length of the loop, and this limiting "average" NEXT power becomes proportional to $f^{1.5}$.

Attenuated or Amplified NEXT (ANEXT). In the CO the source and sink for NEXT are co-located, but at the remote end they may be significantly separated, as shown in Figure 3.10. It can be seen that if the interferer's loop is either longer or shorter than the interferee's, the NEXT will be attenuated by the difference in lengths. For most loop plants the variation of loop length within one binder group is relatively small ($< 20\%$ of the total loop length), and the attenuation of the NEXT is insignificant; the average NEXT is typically only about 1 dB less than the modeled worst-case value.

The effect can be much greater, however, if the interferer is T1,[23] which is repeatered approximately every 3 kft, as shown in Figure 3.11. Then the NEXT

---

[22] "Large" means a loop length for which the attenuation at the frequency of interest is greater than about 20 dB; this would mean that an average propagated, coupled, and returned signal could add only 1% to the NEXT. For ADSL and VDSL, all loop lengths of interest are "large" over the frequency ranges of interest, but this is not true for HDSL2, which tries to use a band almost to dc.
[23] It will occur with any repeatered signal, but T1 is the most virulent.

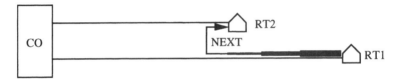

Figure 3.10 Attenuated NEXT from remote transmitter 1 into remote receiver 2.

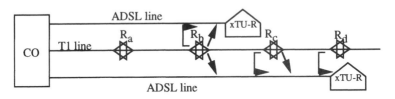

Figure 3.11 NEXT and FEXT from T1 repeaters.

TABLE 3.3 NEXT Losses at 0.772 MHz

	Pair-to-Pair Losses (dB)	49 Pairs-to-Pair Losses (dB)
Median	77	50
1% worst-case	53	45

depends very strongly on the position of the remote unit[24] relative to the repeaters. The average NEXT coupling coefficient is defined in T1.413 as 5.5 dB lower than the worst-case values given in Table 3.3. This calculation was based on the assumption (widely accepted by T1E1.4 at the time) that there may be repeaters both before (upstream) and after (downstream) the xTU-R (e.g., R_b and R_c, respectively, in Figure 3.11). I have heard reports, however, that some LECs do not have repeaters in the distribution cable; if this is so, T1 NEXT would be significantly reduced.

3.6.2 FEXT

The simplest form of FEXT, called *equal-level FEXT* (EL-FEXT), is shown in Figure 3.12(*a*). Transmitters Xmit_{12} and Xmit_{34} and receivers Rec_{12} and Rec_{34} are co-located. It can be seen that all contributions to the FEXT received in Rec_{34} pass through the same length of cable—first in some length of pair 12 and then in a complementary length of 34—as the receive signal itself. That is, if there is no

---

[24] Called xTU-R in the figure to emphasize that all DSL sysems (H, A, and V) are susceptible to T1 NEXT.

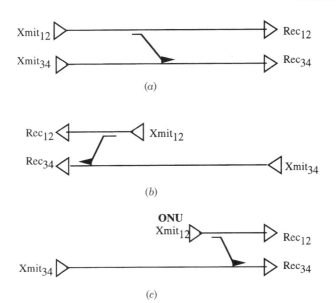

Figure 3.12 FEXT: (*a*) equal level; (*b*) unequal level upstream from close RT; (*c*) unequal level downstream from ONU in cabinet.

interaction between the transfer and cross-coupling functions,

$$H_{\text{FEXT}}(l,f) = \int_0^l H_{12}(\lambda,f)H_{\text{XT}}(\lambda,f)H_{34}(l-\lambda,f)d\lambda \qquad (3.35)$$

Then, by the same simplifying steps as were used for NEXT,

$$H_{12}(\lambda,f)H_{34}(l-\lambda,f) = H(l,f) \qquad (3.36)$$

$$H_{\text{FEXT}}(l,f) = H(l,f)\,f\int_0^l H_{\text{XT}}(\lambda)d\lambda \qquad (3.37)$$

The cross-coupling function $H_{\text{XT}}(\lambda)$ is a random function[25] of λ, and contributions to the integral add only on a power basis. Therefore, the FEXT power transfer function can be written as

$$|H_{\text{FEXT}}(l,f)|^2 = |H(l,f)|^2 k_{\text{FEXT}} l f^2 \qquad (3.38)$$

where k_{FEXT} is an aggregate coupling coefficient that will be different for every pair of pairs.

---

[25] See Section 3.6.4 for more discussion of this "randomness".

EL-FEXT. If the transmitters transmit at the same power level, the signal/FEXT ratio (SFR), which is also called EL FEXT loss, for either pair is

$$\text{SFR}(l,f) = \frac{1}{k_{\text{FEXT}}\, l,\, f^2} \tag{3.39}$$

The arrangement in Figure 3.12(a) is symmetrical; the transmitters could be replaced by receivers, and vice versa, and the FEXT consequences would be the same. All the test loops defined in T1.413 [ANSI, 1995] assumed this situation. This was done for simplicity of testing, but it is a rather unreasonable assumption, for three reasons:

1. It is very unlikely that the loop under test and the interfering loop will have the same arrangement of bridge taps. A bridge tap in only the test loop will reduce the SFR below that given by (3.39); a bridge tap in only the interferer will increase the SFR. Since most bridge taps are closer to the customer premises than to the CO (i.e., in the distribution cables rather than the feeder cable), this effect is more important in the upstream direction.
2. All customer premises are not equidistant from the CO/ONU. If the upstream transmitter Xmit$_{12}$ in Figure 3.12(b) transmits at full power, its FEXT into Rec$_{34}$ will be very severe.
3. If ADSL signals from a CO share a distribution cable with VDSL signals from an ONU, as shown in Figure 3.12(c), the FEXT from a VDSL downstream transmitter Xmit$_{12}$ into an ADSL receiver Rec$_{34}$ may be severe.

Unequal-Level FEXT (ULFEXT). A generalization of (3.39) that takes account of the different loop lengths[26] is

$$\text{SFR} = \frac{|H_1|^2}{|H_2|^2} \cdot \frac{1}{k_{\text{FEXT}}\, l_{\min}\, f^2} \tag{3.40}$$

where H_1 and H_2 are the transfer functions of the signal and FEXT paths, respectively, and $l_{\min}$ is the shorter of the two lengths l_1 and l_2. The effects of ULFEXT on capacities and service offerings are discussed in Section 4.5.

Amplified FEXT. Another type of ULFEXT arises with T1, as shown in Figure 3.11. The outputs of each amplifier must be treated as separate interfering signals that couple through only a section of the loop (3 kft or less), and are attenuated by only a part of the full loop. In this case FEXT is still lower than NEXT, but only by a few decibels; it cannot be ignored.

---

[26] The situations where signal and crosstalker encounter different bridge taps have to be dealt with individually.

3.6.3 Measurements and Statistical Models of Crosstalk

In the 1970s and early 1980s, many hundreds of thousands of measurements of crosstalk were made, mostly on 50-pair binder groups that are used for interexchange transmission[27] of T1 and T2 signals; many measurements were reported in [Lin, 1980]. They were of two types:

1. *Pair-to-pair measurements.* For an N-pair ($N = 50$) cable these involved $N(N-1)/2$ measurements per cable at each frequency of interest.
2. *($N-1$) pairs-to-one pair measurements.* These involved measurements with all interfering pairs simultaneously, independently, and randomly driven, resulting in N power sums per cable.

From both sets of measurements, cumulative probability density functions (CDFs) were plotted, and various worst-case probabilities estimated. For T1 and T2 transmission, which uses multiple repeaters and requires a very high level of quality assurance, the 0.1% and even 0.025% worst-case probabilities were needed, but for all other, unrepeatered, DSL services, the 1% worst-case probability is adequate.

NEXT at 0.772 MHz and FEXT at 3.15 MHz. Lin's results showed that the pair-to-pair losses for both NEXT and FEXT are approximately lognormally distributed (i.e., normally distributed on a log scale) out to the 1 pecentile.[28] His results, rounded to integer dB values, for NEXT (his Figures 6 and 10) and FEXT (his Figures 5 and 9) are summarized in Tables 3.3 and 3.4. These were partially confirmed, with a smaller sample size, in [Valenti, 1997].

1% Worst-Case NEXT. [Unger, 1985] proposed that the 1% worst-case NEXT powers can be accurately represented by

$$10 \log|H_{\text{NEXT}}(l,f)|^2 = -66 + 6 \log(f) \quad \text{dB} \quad \text{for } f < 0.02 \text{ MHz} \quad (3.41a)$$
$$= -50.5 + 15 \log(f) \quad \text{dB} \quad \text{for } f \geq 0.02 \text{ MHz} \quad (3.41b)$$

TABLE 3.4 FEXT Losses at 3.15 MHz

	Pair-to-Pair Losses (dB)	49 Pairs-to-Pair Losses (dB)
Median	65	38
1% worst-case	42	32

---

[27] For the feeder and distribution cables that are used for ADSL and VDSL, 25-pair binder groups are more common.

[28] His main point was to show that a gamma function distribution is a much better fit beyond 1 percentile, but that need not concern us here.

and

$$10\log|H_{\text{NEXT}}(49, f)|^2] = -59.2 + 4\log(f) \quad \text{dB} \quad \text{for } f < 0.02\,\text{MHz} \quad (3.42a)$$
$$= -42.2 + 14\log(f) \quad \text{dB} \quad \text{for } f \geq 0.02\,\text{MHz} \quad (3.42b)$$

The increase from 1 to 49 interferers at 20 kHz is 10.2 dB, but at 0.772 MHz it is 8.6 dB, which agrees well with Lin. For T1.413 the slight difference in slope above 20 kHz was ignored, and a general formula for N interferers was adopted:

$$10\log|H_{\text{NEXT}}(N, f)|^2 = -50.6 + 6\log(N) + 15\log(f) \quad \text{dB} \qquad (3.43)$$

An easily remembered approximation is -40 dB for a fully excited cable at 1 MHz.

NOTE: For a fully excited 50-pair cable, equation (3.42) is about 2 dB pessimistic (i.e., it predicts too high a level of NEXT) compared to Lin's measurements.

[Huang and Werner, 1997] and Zimmerman[29] argued that with a binder group full of interferers, the 1% worst-case crosstalk level is independent of the number of pairs. They modified (3.43) to

$$10\log|H_{\text{NEXT}}(N,f)|^2 = -40.3 + 6\log\frac{N}{N_{\text{bg}} - 1} + 15\log(f) \quad \text{dB} \qquad (3.44)$$

where N_{bg} is the number of pairs in the binder group. This matches (3.43) for a 50-pair cable, but predicts 1.9 dB and 4.4 more NEXT for 25 and 10-pair cables (typically, used only for distribution). It is not clear whether this modification will be accepted.

1% Worst-Case FEXT. The generally accepted formula for EL-FEXT is

$$10\log|H_{\text{NEXT}}(N,f,l)|^2 = -51.1 + 6\log(N) + 20\log(f) + 10\log(l) \quad \text{dB}$$
$$(3.45)$$

where l is in kft. The Werner–Zimmerman modification is

$$10\log|H_{\text{NEXT}}(N, f, l)|^2 = -41.0 + 6\log\frac{N}{N_{\text{bg}} - 1} + 20\log(f) + 10\log(l) \quad \text{dB}$$
$$(3.46)$$

---

[29] Private correspondence.

An easily remembered approximation is $-40\,$dB for 1 kft of fully excited cable at 1 MHz.

Average NEXT. As we shall see in Chapter 4, capacity is proportional to the log of the SNR. Therefore, the most appropriate "average" to use is the median on a log scale; it is the 50 percentile of performance. The median losses are given in Tables 3.3 and 3.4 for pair to pair and 49 pairs-to-pair; the question is how to interpolate for other numbers of interferers. This is a very complicated problem that, as far as I know, has not been solved. For lack of anything better, I propose the same formula as used for the 1% worst-case values: that is, $K_1 + K_2 \log(N)$, with K_1 and K_2 being chosen to match the $N = 1$ and 49 values, with the caveat that it may be several dB in error for the commonly used numbers of 10 and 24 interferers. That is, the median NEXT loss at 0.772 MHz would be approximated by $(77 - 16 \log(N)\,$ dB.[30]

Unfinished Business. The multiplier of 16 for log(N) for average NEXT is counterintuitive; it would seem that the "average" power would be proportional to the number of interferers. The median on a log scale, however, *appears* to increase more rapidly with *N*.

Average FEXT. By the same argument as above, the median *N*-pair-to-pair FEXT loss at 3.15 MHz can be approximated by $[65 - 16 \log(N)]$ dB.

Summary. NEXT and FEXT, average and 1% worst-case values, can all be defined by one expression with four variables:

$$10 \log|\text{XT}|^2 = K_1 + K_2 \log(N) + K_3 \log(f) + K_4 \log(l) \qquad (3.47)$$

where *l* is in kft and K_1, K_2, K_3, and K_4 are defined in Table 3.5.

Plots of NEXT and FEXT. Figure 3.13 shows $|H(f)|^2$ and both 1% worst-case and average $|H_{\text{FEXT}}(f)|^2$ and $|H_{\text{NEXT}}(f)|^2$ values from (3.44) as a function of

TABLE 3.5   Crosstalk Coefficients for Equation (3.45)

	NEXT (dB)		FEXT (dB)	
	Average	1% Worst-Case	Average	1% Worst-Case
K_1	−75	−51	−75	−51
K_2	16	6	16	6
K_3	15	25	20	20
K_4	0	0	10	10

Note: It is a coincidence that K_1 is the same for NEXT and FEXT; if we used any frequency other than 1 MHz or length other than 1 kft for normalization, it would not be.

---

[30] Note that the 6 log(N) scaling is appropriate only for 1% worst-case losses.

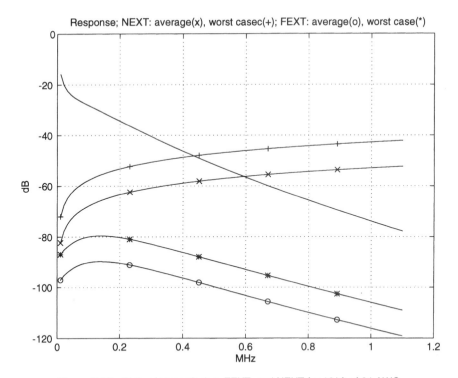

Figure 3.13   Plots of transmission, FEXT, and NEXT for 12 kft of 24 AWG.

frequency for 24 interferers in a 12-kft loop of 24 AWG (typical for ADSL), and Figure 3.14 shows the same for a 3-kft loop (typical for VDSL).

Quadded UTPs. Cabling in Japan and some European countries combines two UTPs in a quad, and then arranges quads in a cable in such a way that the relative positions of the quads is approximately the same throughout the cable. There are therefore three levels of XT coupling coefficients (both NEXT and FEXT): within a quad, between nearest-neighbor quads, and between separated quads.

The mixed model proposed for evaluation of modems on Japanese cables comprises one same-quad, four nearest-quad, and four separated-quad, interferers (i.e., N, the number of interferers $= 9$). The values of K_1 defined in equation (3.47) are given in Table 3.6. For the evaluation of modems on European cables it was decided that there is too much variation in cable characteristics between countries for a common crosstalk model to be possible, and Annex H of T1.413 and G.996 define a simple piecewise-linear noise PSD.

3.6.4   Crosstalk from Mixed Sources

If $N+1$ pairs in a binder group are carrying the same type of signal, the 1 percentile crosstalk power multiplier, $W(N)$, is, as shown in (3.43) and (3.45),

TABLE 3.6   Worst-Case Crosstalk Coefficients for Japanese Quadded Cable

Dielectric	K_1 NEXT (dB)	K_1 FEXT (dB) [normalized to 1 km (1 kft)]
CCP	-45.9	$-42.3\ (-47.5)$
Paper	-43.7	$-34.3\ (-39.5)$

Note: These numbers are considerably higher than for U.S. cables. This may be because one pair is the nearest neighbor for most of the cable and dominates the crosstalk.

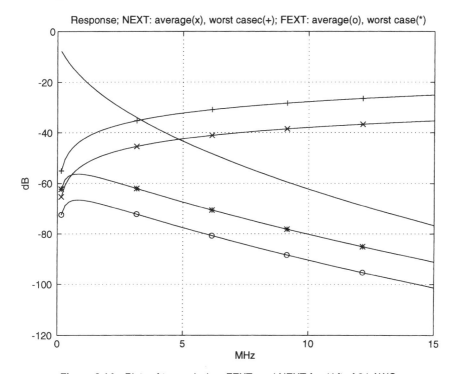

Figure 3.14   Plots of transmission, FEXT, and NEXT for 4 kft of 24 AWG.

$N^{0.6}$. If, however, there are N_1 of one type of signal and N_2 of another type, what are appropriate weightings, $[W(N_1), W(N_2)]$? Clearly,

$$[W(N_1), W(N_2)] \neq [N_1^{0.6}, N_2^{0.6}]$$

because the two sets of pairs cannot simultaneously both be "worst case." If the N_1 are considered to be the worse of the two "worst cases," it was agreed by T1E1.4 that

$$[W(N_1), W(N_2)] = [N_1^{0.6}, \{(N_1 + N_2)^{0.6} - N_1^{0.6}\}] \qquad (3.48)$$

with an obvious extension to any number of ordered interferers.

To avoid the problem of having to define the hierarchy of interferers the Full Service Access Network[31] (FSAN) study group recommended that for N_1, $N_2, \ldots, N_m$ crosstalk signals $S_1, S_2, \ldots, S_m$ (appropriately weighted for PSD and frequency), the total crosstalk should be

$$XT = (N_1 S_1^{1/0.6} + \cdots + N_m S_m^{1/0.6})^{0.6} \qquad (3.49)$$

This reduces to the accepted expressions if $N_m = 0$ and if $S_m = S_1$, but there is still an implicit hierarchy in that the signals with the larger PSDs appear to be the most tightly coupled. As an extreme example, if S_1 is the only nonzero crosstalk signal, then (3.49) reduces to the expression for N_1 *worst-case* interferers. It is thus a conservative (pessimistic) model of crosstalk.

3.6.5   Modeling and Simulation of Crosstalk

CAVEAT: Many of the ideas in this section have not been subjected to peer review; some may be irrelevant, some may even be wrong. They are also incomplete, because no satisfactory model for the variation of the relative positions of two pairs has been found. Please use the ideas warily.

As suggested earlier, the statistical models of crosstalk in the preceding sections are adequate for the rate/range calculations needed for system planning and the design of first-generation DSL modems. There are, however, several reasons why more precise modeling of pair-to-pair crosstalk functions[32] may be needed.

- As DSP gets cheaper, smaller, and less power-consuming, crosstalk cancellation will become more attractive; knowledge of the impulse response of each pair-to-pair path will become important.
- The longest part of the training time of a DSL modem is used for noise estimation (see Section 5.2), and attempts to reduce this typically rely on some assumed frequency-domain correlation of the noise. Plots of pair-to-pair NEXT versus frequency indicate that there may be, in fact, very little correlation!
- Intellectual curiosity demands it.

In this section we discuss how the modeling might be done.

The touchstone of any simulation method is how well the results match reported measurements. The most relevant measurements are those of probability distributions of pair-to-pair and 49 pair-to-pair crosstalk, shown in Figures 5, 6, 9, and 10 of [Lin, 1980], and of crosstalk versus frequency, shown in Figures 4 and 5 of [Huang and Werner, 1997] and Figures 3 and 5 of [Valenti, 1997]. [Unger, 1985] showed the results of a computer simulation of NEXT that

---

[31] A consortium of mostly European and Asian telcos (LECs).
[32] The multiple pair-to-pair numbers are useful only for statistical modeling.

very closely matched the measurements reported in [Lin, 1980], but he described the model only in an unpublished internal Bellcore document,[33] which few outside Bellcore have seen; in its absence we can only begin again!

A mathematical model of two UTPs coupled by distributed capacitances and inductances would be very complicated, and the only feasible approach seems to be to use lumped approximations with small sections of length δl with small lumped coupling capacitances and inductances. Since for any given length of cable, the amount of computation is inversely proportional to δl, this length should be as great as possible. The twist pitch, however, is typically less than 2 ft, and varies from pair to pair, so coupling between any two pairs varies significantly in a few inches. Using $\delta l = 3$ inches would require an enormous amount of computation. The choice of δl should be included in *unfinished business*.

Capacitive Coupling. Figure 3.15(*a*) shows a cross section of a cable with two pairs of wires and four coupling capacitors between the wires. The capacitances and the resultant coupling susceptances, y_{ij}, are dependent on the relative positions of the four wires, which, as can be seen from Figure 3.15(*b*), change continually throughout the cable as the pairs twist (with different pitches) and move around in the cable. The capacitances can be referenced to the self-capacitance of a pair separated by D that was given in (3.1). Hence

$$y_{ij} = j\omega C_{ij} = j\omega C_{\text{per}} \frac{\text{arccosh}(D/d)}{\text{arccosh}(D_{ij}/d)} \qquad (3.50)$$

where D_{ij} is the distance between the centers of the wires.

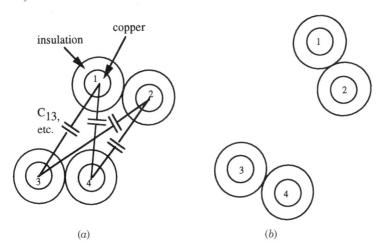

(*a*)   (*b*)

Figure 3.15 Cross section of cable at two different points showing coupling capacitors: (*a*) four coupling capacitors; (*b*) further down the cable: 12 and 34 have separated slightly and have twisted by different amounts.

[33] Maybe this citation will persuade somebody to publish it!

Inductive Coupling. We would like to use the method of calculating the self-inductance of a pair, which was described in Section 3.2, to calculate the coupling inductances[34] M_{ij}. This is questionable, however, because now the current generating the flux linkages may not be in the plane of the loop, and the redistribution of current—with its resultant increase of inductance from the $\ln(2D/d-1)$ form to the arccosh form of (3.2)—is less pronounced. If the redistribution is ignored, Moustafavi has shown[35] that the coupling inductance from wire i to loop jk is

$$L_{i,34} = \frac{\mu_0}{\pi}\ln\frac{D'_{i3}}{D'_{i4}} \qquad \text{for } i = 1, 2 \tag{3.51}$$

where D'_{ij} and D'_{ik} are as defined in Figure 3.16, with the primes used to emphasize that the definition of the distances is slightly different[36] from that used for capacitance in Figure 3.15 and equation (3.50).

NOTE: The $\ln(D/d)$ approximation discussed in Section 3.2 is a special case of (3.51) in which the "intefering" wire becomes one of the pair.

Hence $z_{I,34}$ can be written as a function of L_{per}:

$$z_{i,34} = j\omega\frac{L_{\text{per}}}{2}\frac{\ln(D'_{i3}/D'_{i4})}{\ln[(2D-d)/d]} \tag{3.52}$$

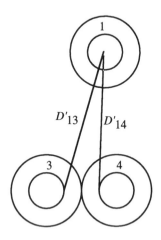

Figure 3.16 Cross section of cable showing D values for calculation of inductive coupling.

---

[34] It is tempting to call these *mutual* inductances, but that would be misleading because it is coupling from one wire into a loop, not into two independent wires.
[35] Private correspondence.
[36] This difference is aesthetically disturbing, but there seems to be no way to avoid it.

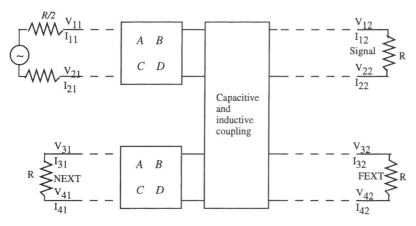

Figure 3.17 One small lumped section of four-port balanced network to model XT.

The reactances $z_{i,jk}$ are shown this way because, as pointed out in [Paul, 1992], current in wire i couples into a loop jk rather than into two independent wires j and k, and the resultant voltages are therefore inherently differential.

Relative Magnitudes of Capacitive and Inductive Couplings. Some discussions of crosstalk ignore inductive coupling altogether, but although for the differential mode it is typically smaller than the capacitive coupling, it is not insignificant.

Matrix Analysis. A differential signal is input to one port of the four-port network shown in Figure 3.17, and the received signal, the NEXT, and the FEXT are delivered to the other three ports. The most straightforward way to analyze this is to consider the voltages and currents in all four wires independently, and discard the common-mode component only when the signals are delivered to well balanced transformers. The coupled two-ports can then be broken into alternating small sections of transmission and coupling, which have chain matrices $\Delta\mathbf{M}_{\text{trans}}$ and $\Delta\mathbf{M}_{\text{coup}}$. If the 8×8 chain matrix of the four-port is initialized as $\mathbf{M} = \mathbf{I}$, then for each lumped element $\mathbf{M}$ can be updated by multiplying from the left by $\Delta\mathbf{M}_{\text{trans}}\Delta\mathbf{M}_{\text{coup}}$. That is,

$$
\begin{bmatrix}
v_{11} \\
i_{11} \\
v_{21} \\
i_{21} \\
v_{31} \\
i_{31} \\
v_{41} \\
i_{41}
\end{bmatrix}
= \Delta\mathbf{M}_{\text{trans}}\Delta\mathbf{M}_{\text{coup}}
\begin{bmatrix}
v_{12} \\
i_{12} \\
v_{22} \\
i_{22} \\
v_{32} \\
i_{32} \\
v_{42} \\
i_{42}
\end{bmatrix}
\tag{3.53}
$$

where

$$\Delta \mathbf{M}_{\text{trans}} = \begin{bmatrix} A & B & 0 & 0 & 0 & 0 & 0 & 0 \\ C & D & 0 & 0 & 0 & 0 & 0 & 0 \\ 0 & 0 & A & B & 0 & 0 & 0 & 0 \\ 0 & 0 & C & D & 0 & 0 & 0 & 0 \\ 0 & 0 & 0 & 0 & A & B & 0 & 0 \\ 0 & 0 & 0 & 0 & C & D & 0 & 0 \\ 0 & 0 & 0 & 0 & 0 & 0 & A & B \\ 0 & 0 & 0 & 0 & 0 & 0 & C & D \end{bmatrix} \tag{3.54}$$

$$\Delta \mathbf{M}_{\text{coup}} = \begin{bmatrix} 0 & 0 & 0 & 0 & 0 & z_{3,12} & 0 & z_{4,12} \\ y_{13}+y_{14} & 0 & 0 & 0 & -y_{13} & 0 & -y_{14} & 0 \\ 0 & 0 & 0 & 0 & 0 & -z_{3,12} & 0 & -z_{4,12} \\ 0 & 0 & y_{23}+y_{24} & 0 & -y_{23} & 0 & -y_{24} & 0 \\ 0 & z_{1,34} & 0 & z_{2,34} & 0 & 0 & 0 & 0 \\ -y_{13} & 0 & -y_{23} & 0 & y_{13}+y_{23} & 0 & 0 & 0 \\ 0 & -z_{1,34} & 0 & -z_{2,34} & 0 & 0 & 0 & 0 \\ -y_{14} & 0 & -y_{24} & 0 & y_{14}+y_{24} & 0 & 0 & 0 \end{bmatrix} \tag{3.55}$$

The resultant matrix with elements $M_{ij}(i,j=1$ to 8$)$ can be condensed to a 4×4 with elements m_{ij} $(i,j=1$ to 4$)$ by considering only the differential-mode components. The result[37] is

$$\mathbf{m} = \begin{bmatrix} 11-13-31+33 & 12-14-32+34 & 15-17-35+37 & 16-18-36+38 \\ 21-23-41+43 & 22-24-42+44 & 25-27-45+47 & 26-28-46+48 \\ 51-53-71+73 & 52-54-72+74 & 55-57-75+77 & 56-58-76+78 \\ 61-63-81+83 & 62-64-82+84 & 65-67-85+87 & 66-68-86+88 \end{bmatrix} \tag{3.56}$$

where, for the sake of brevity, only the subscripts of the m components are shown.

These equations can be solved for the signal, NEXT, and FEXT transfer functions:

$$\frac{v_{12}}{v_s} = \frac{m_{33} + m_{34}G + m_{43}R + m_{44}}{\text{Den}} \tag{3.57}$$

---

[37] This is messy and tedious, and I expect that by now the number of readers has dwindled to almost zero. But exact modeling of crosstalk will be needed, and I can think of no other way to do it; crosstalk *is* just messy and tedious!

$$\frac{v_{31}}{v_s} = \frac{(m_{31} + m_{32}G)(m_{43}R + m_{44}) - (m_{33} + m_{34}G)(m_{41}R + m_{42})}{\text{Den}} \qquad (3.58)$$

$$\frac{v_{41}}{v_s} = -\frac{m_{31} + m_{32}G + m_{41}R + m_{42}}{\text{Den}} \qquad (3.59)$$

where

$$\text{Den} = (m_{11} + m_{21}G + m_{21}R + m_{22})(m_{33} + m_{34}G + m_{43}R + m_{44})$$
$$- (m_{14} + m_{14}G + m_{23}R + m_{24})(m_{31} + m_{32}G + m_{41}R + m_{42}) \qquad (3.60)$$

Unfinished Business: Modeling the Variation of D_{ij} and D'_{ij}. The hardest problem for simulation is how to model the distances between the wires as a function of λ, the distance down the cable. They will vary quite rapidly because of twist of the two pairs, and may, depending on how the cable is made, vary slowly because of relative wander of the pairs in the cable. Possible models include:

- To allow the pairs to twist independently, they are considered to rotate within their own cylindrical tubes, which themselves are packed honeycomb style in the cable. This would be only 50% efficient in its use of cross-sectional area; some compression and violation of "tubular integrity" would certainly occur. It also would not allow for the variation of distance between the centers of the pairs.

- The 100 wires of a 50-pair cable could be considered to be packed—with maximum efficiency—in a honeycomb shape. This would be too rigid: It would not allow for any independent rotation of the pairs or for relative wandering of the pairs.

The first model might be appropriate for tightly twisted pairs such as Cat-5 UTP, but for pairs loosely twisted with a random pitch neither arrangement seems credible. The help of experts from the cable industry is clearly needed.

3.6.6 Discussion of Terminology, and Comparison of NEXT and FEXT

The terms *self FEXT* and *self NEXT* have been used to describe crosstalk from "like" systems, but they are confusing (and maybe misleading). *Self* implies crosstalk from any pair into itself. Strictly speaking, self FEXT for pair 12 of Figure 3.9 would be crosstalk from pair 12 to pair 34, and then back again to 12! This does occur, but its magnitude is a second-order small quantity, and the result becomes indistinguishable from pure propagation on pair 12. Similarly, self NEXT is insignificant, and would, similarly, be indistinguishable from echo. Until alternative terms have been agreed upon I will use *kindred* and *alien* to describe crosstalk from like and unlike systems, respectively.

The term *self NEXT* is also used by those designing dual-duplex modems for LANs to mean NEXT from the other pair in a quad; this is carrying known transmit data, and the NEXT can be canceled. This is similar to echo, and the term should be retained with this meaning.

It can be seen that NEXT occurs only if the downstream and upstream signals use the same frequency band(s) and are transmitted at the same time. Kindred FEXT, on the other hand, will always occur if there are two or more xDSL systems of the same type in a cable. If a system design allows NEXT (kindred or alien) to occur, then it will nearly always have much more serious effects than FEXT;[38] in fact, at frequencies above about 150 khz, NEXT would be so serious that simultaneous transmission and reception becomes impractical.

3.7 RADIO-FREQUENCY INTERFERENCE

There are two types of radio-frequency interference (RFI) with which a DSL system designer must be concerned:

1. *Interference from external RF sources into a DSL receiver.* This is called *ingress.*
2. *Interference from a DSL system into an RF receiver.* This is caused by *egress* (radiation) from the loop.

Both are the result of imbalance in the DSL pair. Ingress occurs because a single RF source couples unequally into the two wires, and egress occurs (even from a pair that carries no common-mode current) because the two wires of a pair radiate unequally to a receiving antenna. For the moment we need consider only ingress; specifications to control egress are discussed in Section 10.1.

The worst effects of ingress occur when the RF signal is coupled into a poorly balanced drop wire and appears directly at the input to a remote receiver. This is the situation that has received most attention; it is discussed in detail in Section 10.3. The differential-mode interference is, however, *added to* the line signal, and will propagate in both directions. Its effect at the other end is discussed briefly in Section 10.3.5.

Three sources of RFI may be important for xDSL service. Two are easily identified: amateur ("ham") radio and AM radio; the third can be described only generally as "all other sources."

Amateur Radio. In the United States ham radios operate anywhere in one of the nine bands shown in Table 3.7. Ham transmitters are single-sideband with a bandwidth of less than 4.0 khz (typically closer to 2.5 khz). During a session they may change carrier frequency, but usually stay within any one band. An excellent discussion of those characteristics of ham radios that are relevant to

---

[38] The exception is FEXT from repeatered T1.

TABLE 3.7   Amateur Radio Bands

Start (MHz)	Stop (MHz)
1.81	2.0
3.5	4.0
7.0	7.3
10.1	10.15
14.0	14.35
18.068	18.168
21.0	21.45
24.89	24.99
28	29.1

Note: The last four are way outside the band proposed for any xDSL use, but they are important for the design of any analog front-end filters (see Section 10.4.4).

VDSL systems is given in [Hare and Gruber, 1996]. ANSI committee T1E1.4 and ETSI committee TM6 have considered many contributions on the subject of the level of ingress to tip-and-ring caused by a worst-case (closest and highest power) ham transmitter (e.g., see [Bingham et al., 1996a]). Their conclusions, which take account of the anticipated balance of drop wires and have been incorporated into the VDSL systems requirements [Cioffi, 1998], are:

Worst-case common mode $+30\,\text{dBm}$

Worst-case differential mode $0\,\text{dBm}$

AM Radio. The AM radio band extends from 0.55 to 1.6 MHz, which overlaps both ADSL and VDSL bands. The signal can be modeled as a fixed-frequency carrier 30% amplitude modulated by a white-noise source bandlimited to 5 kHz. The requirements documents define many different interference situations; two of the worst are $f_c = 0.71$ and 1.13 MHz, with common- and differential-mode signals at $+30$ and $-30\,\text{dBm}$, respectively.

AM Radio and ADSL. ANSI standard T1.413 does not specify any AM radio interference in its main body, which is concerned with North American operation. The arguments seem to have been:

- Close proximity to an AM transmitter is unlikely.
- The balance of a drop wire is better (40 to 60 dB) at the lower AM frequencies.
- In contrast to ham transmission, which is bursty, AM transmission is continuous, so ingress can be allowed for during the initialization.

T1.413 Annex H does, however, define a Noise Model A for European use, which includes 10 single-tone interferers at $-70\,\text{dBm}$ each. ITU SG 15 WP1 may also consider AM radio as a definable impairment, to be included in the test suite in G.996 (see Appendix B).

AM Radio and VDSL. The VDSL systems requirements document, however, does—and presumably the eventual VDSL standard(s) will—specify test levels for AM radio ingress. At 300 ft from a transmitter (a high-density urban environment) common- and differential-mode levels as high as $+30$ and -30 dBm, respectively, are predicted. The assumption of 60 dB of balance at frequencies up to 1.6 MHz is probably far too optimistic, particularly if flat drop wire such as BT's DW8 is used.

All Other Sources. There are, undoubtedly, many other potential sources of RFI—certainly transient, probably bursty. No committee has, however, succeeded in characterizing them, and they are generally lumped together under the amorphous heading of "impulse noise."

4

DSL SYSTEMS:
CAPACITY, DUPLEXING,
SPECTRAL COMPATIBILITY,
AND SYSTEM MANAGEMENT

Capacity and methods of duplexing [i.e., echo cancellation, frequency-division duplexing (FDD), and time-division duplexing (TDD)] are discussed together here because for a crosstalk-dominated medium such as DSL, the capacity in one direction is very dependent on what is being transmitted in the other direction. The sum of downstream plus upstream maximum data rates is probably the best measure of capacity; in Sections 4.3 and 4.4 we describe how to maximize this and partition it between downstream and upstream.

In Section 4.5 we discuss spectral compatibility, which is essentially a way of controlling alien crosstalk. Without putting too much import on the meaning of *compatible*, two different systems A and B might be said to be fully compatible if A interferes with B no more than B does with itself, and vice versa. In Section 4.6 we discuss the impact of spectral compatibility and several other factors on management of a multipair physical layer xDSL system.

4.1 CAPACITY

The term *capacity* as used in this book is a measure of the data rate that can be transmitted through a channel, but it is not the theoretical limit developed by Shannon; it is a practical rate that depends not only on the signal/noise ratio, but also on the method of modulation and demodulation, the coding, the margin, and the required error rate. Each of the latter four factors deserves discussion.

4.1.1 Modulation and Demodulation

For any given channel—and more specifically, for any given channel impulse response (IR)—there are three different upper bounds for performance for single-carrier systems:

1. *Matched filter bound.* This is applicable if a symbol is received in isolation, uncontaminated by previous symbols.

2. *Maximum likelihood bound.* This is achieved if the receiver uses some type of maximum likelihood sequence detection (MLSD), usually a Viterbi detector, to take advantage of the interference from previous and subsequent symbols. If the channel is not too distorted and the IR is not too dispersed, the Viterbi detector can use all the energy in an IR for detection,[1] and the maximum likelihood bound is equal to the matched filter bound.

3. *Decision feedback equalizer (DFE) bound.* This is achieved when the receiver, after appropriate minimum mean squared error (MMSE) prefiltering, subtracts the effects of previous decisions and bases its decision only on the main sample of the IR (defined as h_0).

Only the last of these, the DFE bound, concerns us here because (1) its performance can be directly related to that of a multicarrier system, and (2) the other two detection methods have not been considered for xDSL.[2]

4.1.2 Coding

Two methods of coding are typically used for DSL: Reed–Solomon forward error correction (R-S FEC) and trellis code modulation (TCM). A few details of the separate methods are given in Sections 8.2.4 and 8.2.7, but an analysis of the combination of the two would be extremely complex. For the moment we use crude estimates of the coding gains of the methods defined in T1.413: 2 dB (coding gain = 1.58) for the FEC alone, and 4 dB (coding gain = 2.51) for FEC plus TCM.

4.1.3 Margin

DSL service providers recognize that all measurements of loop impairments are merely samples, and the reported results are statistical. Therefore, to guarantee a particular service to their customers, they require that data and error rates be achieved with all anticipated crosstalk and noise levels increased by some margin m. For ADSL and VDSL 6 dB ($m = 4$) is the accepted value.[3]

---

[1] In [Bingham, 1988] I coined the term *compact* to describe such an IR, but it did not seem to catch on.

[2] A lot of work has been done on precoding to make a multicarrier signal "fit" a distorted channel (see [Ruiz, 1989] and [Ruiz et al., 1992] and references therein). It is not clear (to me, anyway) whether such "vector coding" methods achieve the maximum likelihood bound of a channel.

[3] This is a very pessimistic strategy because, as we saw in Section 3.6, all crosstalk levels are already specified as 1% worst case. There was much discussion in T1E1.4 as to whether *all* noise powers must be quadrupled. The "play-it-safe" conclusion was that they should be!

4.1.4 Error Rate

Opinions about a tolerable bit error rate (BER) range from 10^{-12} for high-quality compressed video to 10^{-4} for data transmission over ATM. For the present discussion we use the number specified in T1.413: 10^{-7}.

4.1.5 The DFE Bound

The probability of a symbol error[4] in the detector of a DFE for a square-constellation QAM signal that conveys b bits (b even) is

$$P_e = 4kQ\left(\sqrt{\frac{3h_0/\mu^2}{2^b - 1}}\right) \tag{4.1}$$

where

$$Q(x) = \frac{1}{\sqrt{2\pi}}\int_x^\infty e^{-y^2}dy \tag{4.2}$$

b is the number of bits per symbol (i.e., 2^b is the number of points in the constellation), k an edge-effect correction factor that approaches unity for large b, h_0 the first (main) sample of the IR, and μ^2 the variance of the noise at the detector.

Equation (4.1) is only approximate for b odd, but it is an acceptable approximation for $b \geqslant 5$ and for calculations of overall capacity; for this purpose $k = 1$ is also an acceptable approximation. For calculations of bit loading (see Section 5.3) the cases of $b = 1$ and 3 must be treated separately. Then the simplified form of (4.1) can be solved for b:

$$b = \log_2\left\{1 + \frac{3h_0/\mu}{[Q^{-1}(P_e/4)]^2}\right\} \tag{4.3}$$

[Price, 1972] showed how for a DFE, (4.3) can be modified to be a function of the SNR at the input to the receiver. Then the data rate for a QAM signal in a Nyquist band from f_1 to f_2 with a margin mar and a coding gain cg is

$$R = \int_{f_1}^{f_2} \log_2\left\{1 + \frac{3\mathrm{SNR}(f)}{(\mathrm{mar/cg})[Q^{-1}(P_e/4)]^2}\right\}df \tag{4.4}$$

where the limits of the band are such that $\mathrm{SNR}(f) \geqslant 1$ for $f_1 < f < f_2$. We will see in Section 5.2 that this expression is applicable, with only minor

---

[4] When using a byte-organized R-S FEC, symbol errors containing only a few bit errors are corrected with the same efficiency as bit errors.

modification, to multicarrier as well, so from now on we use it for capacity calculations.

4.2 DUPLEXING METHODS

The efficiency of a duplexing scheme can be defined as

$$\varepsilon = \frac{\text{total}(\text{down} + \text{up}) \text{ data rate}}{\text{capacity}} \tag{4.5}$$

We consider three methods of duplexing and the efficiencies they can achieve.

4.2.1 Terminology

Before we consider echo canceling (EC) and frequency-division duplexing (FDD) we must clarify what these terms mean: as descriptions of both duplexing *strategies* and/or implementational *tactics*.

- "Pure" EC (Section 4.2.2) means the simultaneous use of the entire band for transmission in both directions; it defines both the strategy and the tactics. It is used in HDSL.
- "Pure" FDD (Section 4.2.3), strictly speaking, refers only to the strategy; it is one of the options for ADSL. It is also often used to describe the tactics of separating the bands by filtering; this, however, can be misleading because Zipper, proposed for VDSL (see Section 10.4) is FDD without filtering.
- EC/FDD (Section 4.2.4) is a mixture: duplex transmission up to some frequency and simplex with FDD above that. It is used in HDSL2, which divides the band into three parts: both/downstream/upstream. The other option for ADSL—EC up to 138 kHz and only downstream above that— is a special case of EC/FDD.
- FDD(EC)[5] (Section 11.4) uses FDD as a duplexing strategy but uses EC instead of filtering to separate the signals.

4.2.2 Echo Canceling

The most efficient way to use any bidirectional medium would seem to be to use EC to remove the reflection of the locally transmitted transmit signal, and transmit in both directions simultaneously using the full available bandwidth for both. For voiceband modems where the noise is (or at least is assumed to be) independent of any other simultaneous transmissions—what we are calling alien noise—this is the best strategy, and an efficiency of 100% can be achieved.

---

[5] This has been called *pseudo-FDD*, but that is misleading because it really is FD*Duplexing*.

Simultaneous transmission in both directions on a DSL, however, causes kindred NEXT; this may require a different approach to duplex transmission. Figure 3.13 shows the levels of signal and kindred NEXT for 12 kft of a 25-pair cable of 24 AWG loaded with 24 kindred systems—a typical medium ADSL loop—and Figure 3.14 shows the same for 3 kft—a typical VDSL loop. The SNR that is needed for transmission depends on the modulation method, the margin, and the coding gain, but a value of 10 dB at the band edge is a reasonable one to use for our first discussion. It can be seen that the signal to NEXT ratios are 10 dB for the two loops at approximately 0.32 and 1.8 MHz. Clearly, simultaneous duplex transmission beyond these frequencies would be impossible; whether it would even be desirable below those frequencies remains to be seen.

4.2.3 Frequency-Division Duplexing

Frequency-division duplexing (FDD) is very well established and understood, so not much needs to be said. The frequency-domain efficiency is

$$\varepsilon_{fd} = \frac{\text{downstream bandwidth} + \text{upstream bandwidth}}{2 \times \text{total channel bandwidth}} \tag{4.6}$$

which, with practical filters, is typically limited to about 40%. The data efficiency, as defined by (4.5), may be slightly more or less than this, depending on the variation of SNR across the band.

4.2.4 EC/FDD

Figures 3.13 and 3.14 suggest that a mixed strategy—duplex transmission using EC up to some crossover frequency, and simplex transmission using FDD above that—might be best. Figure 4.1 shows the aggregate (downstream plus upstream) data rates as a function of this frequency that are achievable (assuming perfect band-separating and perfect EC[6]) on 12 kft of 24 AWG under two different 1% worst-case noise/crosstalk conditions:

1. With 20 ADSL systems as crosstalkers
2. With 10 ADSL systems and 10 HDSL systems[7] as crosstalkers

It can be seen that if there is a mixture of ADSL and HDSL in the binder group, the band up to about 200 kHz is already degraded by the full-duplex HDSL, so the ADSL might as well use it similarly: that is, EC up to about 200

---

[6] Neither is realistic of course, but we are interested here only in comparisons, not in absolute performance.
[7] HDSL is an average interferer: not as bad as T1 and worse than BR ISDN; see Annex B.2 of T1.413 for a definition of its PSD.

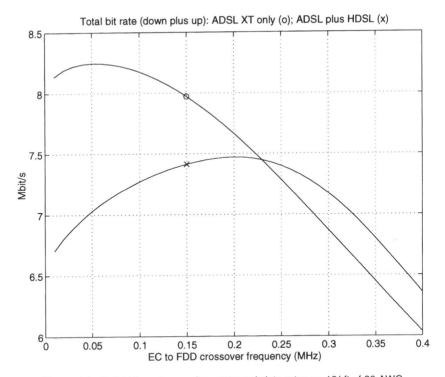

Total bit rate (down plus up): ADSL XT only (o); ADSL plus HDSL (x)

Figure 4.1   Total (downstream plus upstream) data rates on 12 kft of 26 AWG.

kHz and FDD or TDD beyond. This is the strategy that was adopted (somewhat serendipitously!) for ADSL in T1.413: EC up to a nominal 134 kHz, and only downstream beyond that. If, on the other hand, there are only kindred systems in the cable, the total capacity is maximized by using echo cancellation up to only about 60 kHz; in fact, the maximum is so broad that the benefits of any EC at all are very small. This is an important point; if there is to be mass deployment of ADSL, it will soon greatly outnumber older systems, and EC will become almost useless.

4.2.5   Time-Division Duplexing[8]

Time-division duplexing (TDD) is, in an imprecise way, the dual of FDD, but it is simpler in that the system design does not depend on complicated filter calculations. A superframe is defined comprising, in sequence, downstream transmission, a quiet or guard period, upstream transmission, and a second quiet period: d/q/u/q. Then the data rate efficiency is equal to the time-domain

---

[8] TDD has also been called time compression multiplexing, but TCM is now the abbrebiation for trellis code modulation. Only TDD should be used from now on.

efficiency:

$$\varepsilon = \varepsilon_{td} = \frac{t_{down} + t_{up}}{2 \times (t_{down} + t_{up} + 2t_{guard})} \tag{4.7}$$

One system constraint is that the guard period must be greater than the one-way propagation delay, t_{prop}, so it might seem that nearly 50% efficiency could be achieved by making the superframe period ($t_{down} + t_{up} + 2t_{guard}$) very long. There is usually, however, a system constraint on the maximum latency,[9] t_{lat} ; the superframe period must be less than this. Therefore, the maximum efficiency

$$\varepsilon_{max} = \varepsilon_{td} = \frac{t_{lat} - 2t_{prop}}{t_{lat}} \tag{4.8}$$

For both ADSL and VDSL systems the specifications of t_{prop} and t_{lat} would allow an implementation of TDD that achieves about 45% efficiency, but nevertheless, EC/FDD was chosen for ADSL.

Early Use of TDD. TDD was selected as the multiplexing technique for BRI in Japan (see Annex III of ITU-G.961 and Section 9.3) and was also considered for Europe.

Synchronized TDD. Because TDD uses the full band for transmission in each direction, kindred NEXT would be fatal; it must be avoided by synchronizing the frames in all the systems in a binder group. This is done by using the same frame clock for all transmitters in a CO or ONU, and loop timing (i.e., frequency and time locking to the downstream signal) at all the remotes (see Section 10.2). This synchronization may be more difficult now that *unbundling*, the leasing of pairs in a cable (probably without the associated frame clock) to a competitive local exchange carrier (CLEC) and/or an Internet service provider (ISP), is becoming common. The pros and cons of synchronized DMT (SDMT) for VDSL are discussed in detail in Chapter 10.

4.3 CAPACITY REVISITED

With the general principles of duplexing established above, it will be useful to calculate downstream and upstream capacities with 10 ADSL and 10 HDSL crosstalkers under three different sets of conditions:

1. *T1.413 guaranteed performance*: that is,
   1.1. All crosstalks at 1% worst-case level
   1.2. EC up to 130 kHz; only downstream above that

---

[9] This is a ridiculous name for a system processing delay, but I have given up fighting!

1.3. Margin $= 6$ dB

1.4. Total coding gain $= 4$ dB (i.e., R-S FEC and TCM)

2. *Same as condition set 1 except that only FDD is used* (upstream up to 120 kHz, downstream from 155 kHz)

3. *Average performance:* that is,

    3.1. Crosstalk at median level

    3.2. FDD as in condition set 2

    3.3. Margin $= 0$ dB

    3.4. Coding gain $= 2$ dB (i.e., no TCM)

The downstream and upstream rates are plotted as a function of length for 24 AWG in Figure 4.2(*a*) and (*b*). It must be emphasized that these figures are for idealized conditions: no bridge taps, front-end noise ≤ -140 dBm/Hz, no VDSL crosstalk. As befits such conditions, the "T1.413" rates at 12 kft are about 15% higher than specified for CSA loop 8.

The new and enticing part of these results is what can be achieved "on average." The high rates at 10 and 12 kft are not particularly interesting, but the more than threefold increase over the conventionally specified system beyond 18 kft is very intriguing. It must be emphasized, however, that for these very long loops and with kindred FEXT as the only significant crosstalk, the total "noise" is dominated by the so-called AWGN, which was optimistically—and many would argue, unrealistically—set at -140 dBm/Hz. Achieving this level of front-end noise will be a severe and crucial challenge for the analog designer, because with a 20-kft loop, each decibel above -140 dBm/Hz costs approximately 150 kbit/s of downstream data rate. One consolation to be derived from the fact that for the long loops the performance is noise limited is that the available data rate does not change much with the number of interferers, thus simplifying the task of dynamic rate adaptation (see Section 4.6.2).

4.4   A DECISION: EC OR NOT?

As we have seen, compared to conventional FDD (downstream above upstream), EC increases the downstream rate by using the low band for transmission but decreases all upstream rates by subjecting signals received at the CO on that pair to uncanceled echo, and on *all other pairs* to kindred NEXT. The frequency range over which the increase exceeds the decrease (for a net increase in duplex capacity) decreases with loop length.

For ADSL applications within the CSA[10] for which (1) the downstream rate is paramount and (2), a lot of alien NEXT is expected, EC in the band up to 138 kHz, as defined by T1.413, was a good choice. On the other hand, for longer

---

[10] For example, video on demand (VoD).

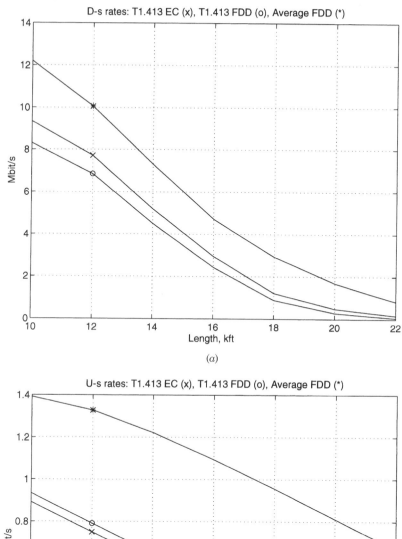

Figure 4.2 Data rates for 24 AWG as a function of loop length: (*a*) downstream; (*b*) upstream.

loops and ADSL[11] applications in which (1) a ratio of upstream to downstream rates of at least 1:8 is needed and (2) ADSL is the majority service in the cable, the disadvantages of EC—reduced upstream rate, much greater complexity, longer training time, and greater difficulty of spectrum management—outweigh its advantages.

Furthermore, as discussed in Section 9.1.3, if the customer-premises unit is operated without the low-pass filter part of the splitter ("splitterless"), it may be necessary to reduce the upstream transmit power in order to avoid noise in the telephone speaker. In this case the signal received at the CO would be even more vulnerable to kindred NEXT and uncanceled echo. EC is perhaps not obsolete, but for all the reasons discussed above it has to be concluded that for *most* of the emerging ADSL market it is obsolescent.

Loops with Very High Noise. On loops with noise according to model B in Annex H of T1.413[12] the usable band for ADSL downstream extends up to only about 400 kHz. Use of the band up to 150 kHz for downstream—made possible by EC—is highly desirable, perhaps even essential.

4.5 SPECTRAL COMPATIBILITY

The (physical layer) system that we should consider here is the combination of a large feeder cable, an FDI, and many distribution cables, shown in Figure 3.1. These together have to deliver a variety of xDSL services to customers over a wide range of distances from the CO (some perhaps < 5 kft, some > 20 kft).

One basic principle of design for xDSL systems is that each new xDSL system should[13] meet its rate/range requirements with crosstalk from kindred systems or any and all previously standardized ("old") systems in the same binder group. I know of only two exceptions to this:

1. T1.413 did not define any tests with ADSL and T1 systems in the same binder group; crosstalk from an adjacent binder group is assumed to be reduced by 10 dB. The implication was that local exchange carriers (LECs) would keep ADSL and T1 separate all the way out to the customer premises.

2. The specifications of ADSL out-of-band PSDs were tightened in Issue 2 of T1.413 to reduce crosstalk into VDSL, a newer system.

---

[11] Only ADSL is considered in this section because the TDD system proposed for VDSL avoids echoes and NEXT completely.

[12] This model is for European loops; it defines the noise as -100 dBm/Hz from 10 to 300 kHz and thus avoids all problems of defining crosstalk. I do not know how realistic it is.

[13] I use "should" rather than the "shall" of standards because the entire subject is still too vague for imperatives.

This principle might be called *spectral optimization*; it is what a lot of the rest of this book is about. The complementary principle of *spectral compatibility* is that each new system should cause no more degradation (via crosstalk) to existing systems than they do to themselves. "No more degradation" is, however, controversial, and three interpretations have been proposed:

1. The PSD of the new system must fall below that of at least one old system at all frequencies. This would guarantee that the new system would cause less crosstalk than would the established system.
2. The PSD of the new system should fall below a superset of old systems; that is, at any frequency a new PSD may be as high as the highest old one.
3. The capacity of the old systems with crosstalk from the new system must be no less than with any mix of old systems [Zimmerman, 1997a and 1997c].

It has been recognized by the spectral compatibility group of T1E1.4 that rule 1 is too stringent; it would unnecessarily inhibit the future development of systems. Indeed, neither ADSL nor the emerging VDSL meets this requirement; their bandwidths are much greater than anything previously used. On the other hand, rule 2 is too lax; Figure 4.3 shows the PSDs of BRI, HDSL, and upstream

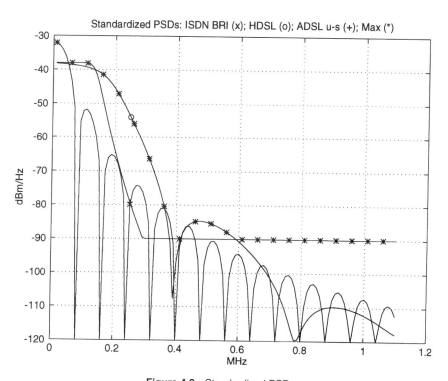

Figure 4.3 Standardized PSDs.

ADSL, and the maxima thereof. A new system could, in aggregate, be a more severe disturber than any old one.

Rule 3 requires the calculation of capacity according to either equation (4.3) for DMT or similar equations given in [Zimmerman, 1998] for PAM, QAM, and CAP systems. It allows the trade-off of higher PSD (and therefore higher crosstalk) in some frequency regions (preferably those where the SNR is already low) for lower in other regions. The first application of this rule was in the PSD of HDSL2,[14] which is shown in Figure 4.4, and it is likely that T1E1.4 will issue a standard based on rule 3 by mid-1999.

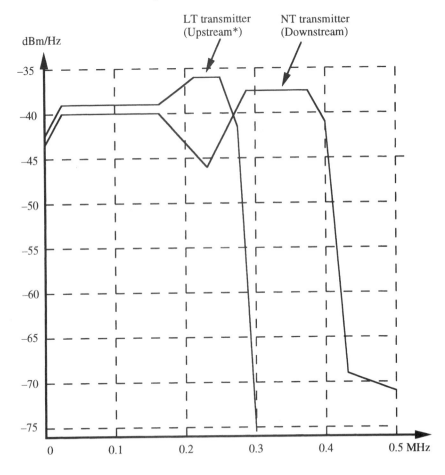

*NOTE: LT and NT (Local and Network Termination) are the usual terms for HDSL2. Downstream and upstream are appropriate only when it is used in the local loop

Figure 4.4 PSD of HDSL2.

---

[14] Such a widely varying PSD in the passband would be a natural for DMT, which can achieve such variations without causing signal distortion. I regret that DMT was not pushed harder for HDSL2.

4.6 SYSTEM MANAGEMENT

Management of the xDSL system shown in Figure 3.1 requires many decisions by a LEC; some of them are:

1. How to offer unbundled pairs to a competitive LEC while maintaining spectral compatibility and other necessities for service quality
2. What mix of data rates and services can be offered, and whether these should be adaptive
3. Whether to impose controls on some transmit PSDs beyond those defined by the standard
4. Whether to enable some of the options defined in the standard (e.g., EC in ADSL)
5. In which binder group to put the pairs carrying different services

These problems are discussed in Sections 4.6.1 through 4.6.5.

4.6.1 Local Exchange Carriers: Incumbent and Competitive

Traditionally, the loop plant has been owned and operated by LECs; these were originally part of Bell Telephone and then of the regional bell operating companies (RBOCs). The LECs owned the plant and provided the services thereon, so the only criterion by which they were judged was the quality of those services. The Telecommunications Act of 1996, however, required that in exchange for permission to compete in the long-distance transmission market, the incumbent LECs (ILECs) must "unbundle" their cables and lease some of their loops to competitive LECs (CLECs), which could provide some DSL services. Section 271 of the act defined the criteria by which it would be judged whether the local market had indeed become competitive.

One of the ways in which compliance with section 271 could be demonstrated is by providing to the CLECs more information about the loops and their data-carrying potential.[15] In ascending order of sophistication, the details of each leased loop that are needed are:

1. Assurance that loading coils and repeaters have been removed.
2. Length and dc resistance: usually readily available.
3. DSL services presently provided and contemplated for all other pairs in both the feeder *and* distribution binder groups. This is discussed in more detail in Section 4.6.5.
4. The ADSL capacity of that loop with some agreed-upon set of interferers. In Section 4.3 we showed how to analyze a loop given its length and gauge and the position, length, and gauge of all bridge taps, and in Section

---

[15] *Capacity* is defined in Chapter 4, but it is too precise a word for here.

4.1.5 we showed how to calculate the capacity. It has been claimed, however [Sapphyre, 1998], that such detailed information of the frequency response is not needed and that capacity can be calculated approximately just from measurements made in the voiceband using a V.34 modem. Whether the method can (and will) be widely used remains to be seen.

Reverse ADSL. If space for its equipment is not available in the CO, a CLEC may use a nearby building, and access the CO via other pairs that are cross-connected in the CO to the pairs going to its customers. This reverse ADSL (transmitting a downstream signal upstream, and vice versa) configuration is shown in Figure 4.5, where the CLEC customer is shown closer to the CO to emphasize that the range is reduced by l_B, the distance of the CLEC's office from the CO. Such an arrangement would, strictly speaking, violate many of the spectral rules discussed in the preceding section, but a pragmatic relaxing of the rules is needed; it can be shown as follows that the crosstalk into orthodox ADSL pairs can be held to an acceptable level.

Figure 4.5 shows two other types of crosstalk that will be generated from the CLEC's "ATU-C" on pair B[16] into pair A: ANEXT into the downstream and, if the CLEC uses EC, UL FEXT into the upstream. A reasonable requirement is that both of these should be less than what conventional ADSL systems to do themselves: that is, kindred FEXT and kinderd NEXT, respectively. The limitations on l_B can therefore be calculated as follows.

ANEXT. In the downstream band

$$15\log(f) - \alpha(l_A - l_B)\sqrt{f} < 20\log(f) + 10\log(l_A) - \alpha l_A\sqrt{f} \qquad (4.9)$$

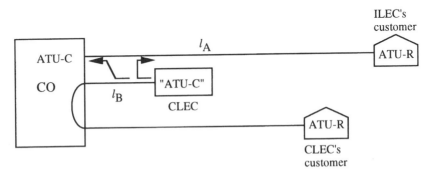

Figure 4.5   Reverse ADSL to access ATU-C outside the CO.

---

[16] The crosstalk from the CLEC's upstream signal going downstream on pair B will be attenuated and inconsequential.

That is,

$$l_B < \frac{5\log(f) + 10\log(l_A)}{\alpha\sqrt{f}} \qquad (4.10a)$$

For 26 AWG, the most common gauge out of the CO, $\alpha \approx 7.8\,\text{dB/kft}$, and the CSA limit for l_A is 9 kft, so that at $f = 0.3\,\text{MHz}$ (the "sweet" part of the downstream band) $l_B < 1.63\,\text{kft}$, and at $f = 1.0\,\text{MHz}$ $l_B < 1.22\,\text{kft}$. By averaging the crosstalks across the whole downstream band a conservative constraint is that

$$l_B < 1.5\,\text{kft} \qquad (4.10b)$$

ULFEXT. At $f = 0.12\,\text{MHz}$, the upper limit of the upstream band, ULFEXT is greatest at approximately 1.5 kft, so it should be compared with kindred NEXT at this frequency and range. Downstream and upstream PSDs are -40 and $-38\,\text{dBm/Hz}$, respectively, so

$$\text{NEXT} = -40 - 51 + 15\log(0.12)$$
$$\approx -105\,\text{dBm/Hz} \qquad (4.11)$$

$$\text{ULFEXT} = -38 - 51 + 20\log(0.12) + 10\log(1.5) - 6.2 \times 1.5 \times \sqrt{0.12}$$
$$\approx -109\,\text{dBm/Hz} \qquad (4.12)$$

Therefore, the ULFEXT constraint on l_B is looser than the ANEXT constraint. Hence, if the reverse distance is limited to less than 1.5 kft, reverse ADSL will present no problems to any conventional systems. Since it is in the CLEC's interest to minimize this distance because it subtracts from the achievable range to their customers, the proposed limit of 1.5 kft should be acceptable to everybody. If the distribution binder-group already contains T1, reverse ADSL from a 1.5 kft distance would generate *much* less NEXT than would the T1, and the 1.5 kft limit could be stretched significantly if the ILEC and the CLEC could agree on a rule.

Extended Reverse ADSL. Another type of reverse ADSL does not use a cross-connect to a conventional pair, but terminates the loop with an ATU-R in the CO. The services offered in this case are totally different, but the NEXT dependence on l_B would be exactly the same as in the preceding case. Conservatively, x should be limited to 1.5 kft, but could be stretched depending on what else is in the binder group.

4.6.2   Mix of Data Rates and Rate Adaptation

One of the big advantages of DMT is its rate adaptability, and basic rate adaptation in the PMD layer—finding the maximum data rate that a particular line at a particular time (startup or any time thereafter) can transport—is relatively straightforward; adaptation at startup and on-line are discussed in Sections 5.3 and 8.5.4, respectively. Rate adaptation in the higher layers is, however, much more complicated. Section 4.3 shows that there is a big difference between maximum guaranteeable and average data rates. This difference gets even bigger if one considers loops that are worse than the sets defined in T1.413. I believe that the biggest challenge in the wide deployment and acceptance of ADSL is at the higher layers; can they deal with (and get the most out of) a medium that may offer a range of data rates as high as 6: 1? Some of the questions that must be considered are:

- Can dynamic rate adaptation (DRA) be used effectively at the higher layers?
- Can all applications in the higher layers accept a short break in transmission during a rate change, or must DRA be restricted for some applications?
- Would a "seamless" method of DRA (i.e., with no errors or interruption of service during a rate change) be more acceptable?
- Even if a high data rate is possible upon startup (because there are no interferers), should the rate be limited to that which can be achieved on that loop under some "almost-worst-case" scenario?
- How much information can the LEC provide about how many *potential* interferers there are in the binder group: that is, about the worst case for that loop?
- How can that information be passed to the ATU-C or VTU-O so that it can be used during startup?

4.6.3 PSD Controls

The usual interpretation of the roles of spectral compatibility and system management is that spectral compatibility should be ensured by standards and that system management should accept the spectra thus defined and work with them. There may, however, be some circumstances under which more system-specific control over the PSDs is desirable. Examples of these are:

- If at any time a receiver determines that the margin for the requested service is unnecessarily high, the performance of the system as a whole would be improved if the transmitter at the other end transmitted at a lower PSD level. Because the receiver sends PSD instructions to the transmitter via the g_i (see Section 8.5.1), it can, if told to do so by a higher layer, reduce the far-end's transmit PSD.
- In a VDSL system, *unequal-level* FEXT (ULFEXT) from a close upstream transmitter can be a serious impairment (see Section 10.1.2 for

more details); if this is not controlled explicitly in the eventual VDSL standard, system management should do so.

4.6.4 Enabling or Disabling Options

There are two types of options that may be enabled or disabled during initialization of a pair of modems: those that can be negotiated by the modems themselves and do not require any systems management (e.g., a feature such as trellis coding would be used if both modems have the capability) and those that require a higher-layer decision (e.g., even if both modems have the capability, is this feature desirable for the system?). We need to consider here only the second type; of the following, the first two are allowed for by T1.413, and the third will have to be anticipated in VDSL:

1. *EC*. The possibility of a mixture of EC and FDD ADSL systems[17] in the same binder group raises the question of what the goal is for the overall system. Is it to maximize the sum of the data rates on all the loops, or to maximize the lowest data rate? There are certainly some cables for which these two goals would be opposing, but most service providers would probably choose the "maxmin" goal. The path to that goal, however, is tortuous. On very long loops (> 20 kft) the downstream rate might be significantly increased by using EC, but the upstream rate could be maintained only if the other ADSL systems in the binder group *did not use EC!*

2. *DRA*. If the *application layer* cannot deal with a physical layer that changes data rate during a session, DRA must be disabled.

3. *Upstream power control*. This is particularly important in a TDD implementation of VDSL, where the upstream signal uses the full frequency band; ULFEXT from a full-power, close-in upstream transmitter would be lethal to other VTU-O receivers.

4.6.5 Binder-Group Management

LECs have very little freedom in populating the binder groups in the distribution (F2) cables going to the customer premises; the mix of services must be determined by customer requirements. On the other hand, they have a lot of freedom in the binder groups in the feeder (F1), cable emerging from the CO; the cross connections in the FDI can be used to transfer pairs as needed.

Segregation Out of the CO. An important question is: Can this freedom to configure the binder groups out of the CO be used advantageously?

---

[17] Kindred certainly, but brothers and sisters often fight!

The choice, basically, is between *binder-group continuity*, in which, as far as possible, pairs leave the CO in what eventually becomes the distribution binder group (thereby minimizing cross connections), and *binder-group homogeneity*, in which the binder groups in the *feeder cable* contain, as far as possible, kindred[18] systems. The choice determines the amount of alien NEXT (aNEXT) at the CO, and so affects the upstream performance.

In the band below 138 kHz the PSDs of ADSL downstream and HDSL are similar (ADSL is about 2 dB higher), so if echo-canceled ADSL systems are allowed, the three arrangements — ADSL only, HDSL only, and mixed — all generate similar levels of NEXT. There is little difference between binder group continuity and homogeneity; management cannot achieve much.

If, however, EC ADSL systems are not allowed, binder-group homogeneity would mean that the upstream ADSL experiences only kFEXT from pairs in the same binder group and aNEXT from pairs in other binder groups. This does not, of course, completely eliminate aNEXT, but reduces it by at least 10 dB, which is the number generally agreed upon for adjacent binder groups. The upstream capacity would be considerably increased, and this will become important as ADSL ranges are stretched and other quasi-symmetrical services are offered.[19]

Some LECs use homogeneous binder groups and, furthermore, try not to put different DSL systems in adjacent binder groups: thereby apparently reducing aNEXT even further. However, when the cable is spliced, which may be only 300 ft from the CO, there is no guarantee that the relative positions of the binder groups in the cable will be maintained; two separated binder groups may become adjacent. The conservative way to estimate upstream capacities is therefore to assume aNEXT at adjacent binder-group levels right at the CO.

Segregation of T1. Below about 200 kHz, NEXT from both downstream ADSL and HDSL is higher than from T1, so it is not necessary to segregate T1 in order to protect upstream ADSL. However, as discussed in Section 3.6.2, AmpFEXT from downstream T1 would be a significant source[20] of crosstalk into downstream ADSL if T1s were not segregated. It appears that T1 systems should be kept in separate binder groups wherever possible: that is, in both the feeder and distribution cables.

4.6.6 Rates, Ranges, or Numbers of Customers?

Modem engineers are typically concerned with rate as a function of loop length and amount of crosstalk, system engineers with range at a given rate, and LEC

---

[18] The upstream capacity of pairs in a homogeneous HDSL binder group would be reduced slightly because of increased kNEXT, but HDSL systems should have been originally engineered to expect the worst-case XT anyway.

[19] For example, a videoconferencing requirement of 384-kbit/s symmetric plus whatever ABR is possible.

[20] And, depending on the location of the ATU-R relative to the repeaters, might even occasionally be the dominant source.

business managers with the percentage of their customers who can be offered a particular service. It is interesting to consider the relation between these three parameters.

HDSL in Same or Adjacent Distribution Binder Group. As an example, consider ADSL downstream transmission on long loops (about 18 kft) with HDSL in the same or in an adjacent binder group in the distribution cable.[21] Table 4.1 shows three ways in which the comparison might be made for long 24-AWG loops. It can be seen that, expressed as a percentage increase from the same to the adjacent binder group,

$$\Delta \text{Rate}(170\%) \gg \Delta \text{Range}(13\%) > \Delta \text{Number of customes}(5\%) \quad (4.13)$$

A 5% increase in the number of potential customers does not increase LEC revenues by much, so by this criterion there would be little incentive to manage the distribution binder groups or to improve the equalizers.

Variation with Crosstalk Conditions. As another example, Table 4.2 shows the variation of achievable downstream rates (on a 16.5-kft loop of 24 AWG) with crosstalk conditions: from no crosstalk (someone who signs on at 3 A.M.?) to the

TABLE 4.1   Improvement in Rate, Range or Percentage of Customers Reached

	Rate at 18 kft (Mbit/s)	Range at 1.544 Mbit/s (kft)	Percentage of Customers Served (note 1)
HDSL in same binder group	0.68	16.5	79
HDSL in adjacent binder group (note 2)	1.84	18.6	83

Notes:
1. From Figure 10.3 of [AT&T, 1982]: these percentages are national averages; they may vary significantly from one LEC to another.
2. With only ADSL FEXT, the performance becomes very dependent on front-end noise level and residual distortion. These numbers assume −135 dBm/Hz noise and nearly perfect equalization: both very optimistic!

TABLE 4.2   Variation of Achievable Data Rate with Crosstalk Conditions

	Rate (Mbit/s)
Noise only	3.18
1 average ADSL crosstalker	3.03
1 average HDSL crosstalker in same binder group	2.33
Full worst-case (10 ADSL + 10 HDSL)	1.544

[21] In a new residential area an LEC might consider offering only ADSL in order to increase the downstream rate and/or range.

full 10 ADSL and 10 HDSL crosstalkers specified in T1.413 (achieving basic 1.544-Mbit/s service).

The Lure of Higher Data Rates. In both cases above it can be seen that a more than 2:1 increase in data rate could frequently be offered *if* the TC and higher layers could be set up to take advantage of them.[22]

4.7.   SPECTRAL MANAGEMENT STANDARD: STATUS, FALL 1999

A draft standard was sent out for letter ballot [T1E1, 1999], but has provoked strong objections. The main controversies are:

- Should this be a standard for spectral management, or should it merely define spectral compatibility, and leave the management responsibility to the FCC?
- *Guarded systems* are defined as those with which all new systems must be "spectrally compatible,"[23] but what systems are guarded: only those that have been standardized, or all systems that have so far been "widely deployed?" Does the category include, for example, repeated HDSL (a favorite of ILECs) and/or various non-standard SDSLs (favorites of CLECs), both of which can interfere with ADSL more severely than any previously considered interferers?

The proposed standard was regarded by some CLECs as too protective of the status quo and restrictive on new and innovative services, and they feared that it would be used by ILECs as a reason (excuse?) for refusing to lease pairs to CLECs.[24] Conversely, it was considered by some manufacturers and operators to be too permissive concerning unstandardized (and therefore uncontrolled) systems. How this will be resolved remains to be seen.

---

[22] *And* the front end and the equalizer can be designed to reduce added noise and distortion to below $-135\,\mathrm{dBm/Hz}$.
[23] Recall that compatibility is bi-directional.
[24] This has been thought and spoken, but perhaps never published before now!

5

FUNDAMENTALS OF MULTICARRIER MODULATION

5.1 BLOCK DIAGRAM

A very simplified block diagram of an MCM transmitter is shown in Figure 5.1; the receiver—at least at the present level of detail—is the mirror image of the transmitter. The input to the S/P converter is a sequence of symbols of B bits each; the output for each symbol is N_{car} groups of $b(n)$ bits each. That is,

$$B = \sum_{n \leqslant N_{car}} b(n) \tag{5.1}$$

Some of the $b(n)$ may be zero, but that need not concern us yet. The groups of $b(n)$ are then constellation-encoded, perhaps filtered, and then modulated onto N_{car} subcarriers; the methods of encoding and modulation are considered in Chapters 6 and 7.

The output of the modulator for the mth block is given by

$$y(mT + t) = \sum_{n \leqslant N_{car}} \text{real}\{p(m, n, t)\exp(j\,2\pi n(t/T)\} \qquad \text{for } 0 < t \leqslant T \tag{5.2}$$

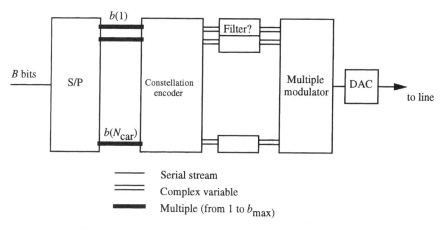

Figure 5.1 Simplified block diagram of an MCM transmitter.

where $p(m,n,t)$ is the baseband pulse resulting from any filtering of the output of the constellation encoder. [Zervos and Kalet, 1989] and [Kalet,1989] showed that if the integration of (4.4) is replaced by a summation over some large but finite set of subcarriers, each with bandwidth δf, then (4.4) applies also to multicarrier modulation. That is,

$$R = \delta f B = \delta f \sum_{n \leqslant N_{\text{car}}} b(n) \qquad (5.3)$$

where

$$b(n) = \log_2\left\{1 + \frac{3\text{SNR}(n)}{(\text{mar}/\text{cg})[Q^{-1}(P_e/4)]^2}\right\} \qquad (5.4)$$

Equations (5.4) can be simplified by combining the terms that are not functions of n into one variable:

$$\gamma = \frac{\text{mar}}{3\text{cg}}[Q^{-1}(P_e/4)]^2 \qquad (5.5)$$

so that

$$b(n) = \log_2\left[1 + \frac{\text{SNR}(n)}{\gamma}\right] \qquad (5.6)$$

For an ADSL system with BER $= 10^{-7}$, mar $= 4$ (6 dB), and cg $= 2.51$ (4 dB), the "gap"[1] $\gamma \approx 14.0$.

Equations (5.3) and (5.6) are a little more restrictive than (4.4) in that the range for n must be narrowed so that the argument of the log is at least 2.0 (i.e., b_n, the number of bits assigned to each subcarrier, $\geqslant 1$), and, furthermore, b_n must be integer. Even though the requirement for single-carrier modulation is only that the SNR should be greater than unity, these extra constraints on multicarrier modulation are fairly inconsequential because:

- The capacity of the edges of the band beyond $b(n) = 1$ and out to SNR $= 1$ is very small, and a decision-feedback-equalized single-carrier modem can make use of those edges only with a large (perhaps impractically large) equalizer.
- We shall see how the $\log_2(\cdot)$ can be rounded to the nearest integer with no increase in power or loss of capacity.

---

[1] So called by [Starr et al., 1999] because it is the gap between the practically achievable and the Shannon limit. For no margin and no coding gain and an error rate of 10^{-7} $\gamma = 9.0$ (9.5 dB).

The most important point about equations (5.3) and (5.4) is that the number of bits assigned to each subcarrier must be calculated from the SNR *and* sent back to the transmitter. This feedback from receiver to transmitter is analogous to the precoding ([Tomlinson, 1971] and [Harashima and Miyakawa, 1972]), that is used in single-carrier systems with severe channel distortion to avoid error propagation in the DFE; this is discussed in more detail in Sections 5.3 and 7.1.1.

By contrast, OFDM systems, which are used primarily for broadcasting—with no feedback possible from receiver(s) to transmitter—use a constant (or at least fixed for a transmission session) bit loading. If this were used for transmission via the DSL, where the SNR varies widely across the band, then either the bit loading would have to be very conservative in order to protect the subcarriers with lowest SNRs, or the error rate on those subcarriers would be very high and would greatly degrade the performance.

5.2 CHANNEL MEASUREMENT

Calculation of the SNR requires two measurements for each subcarrier: of the channel response and of the variance about that response caused by *noise*, which is the sum of conventional noise, crosstalk, and residual (after equalization) channel distortion. The two measurements can be combined by the transmission of a pseudo-random sequence—using all subcarriers[2]—that subjects the transmitter/channel/receiver to all possible distorted sequences. The only question is how many blocks or symbols are needed. There are two requirements:

1. The error in the estimate of the response must be small enough that the "offset" (actual response minus assumed response) does not contribute significantly to the total error during data transmission.

2. The error in the estimate of the SNR must be small enough that it does not significantly affect the bit loading. If the aggregate noise is assumed to be Gaussian distributed, the standard deviation of the estimate of the SNR is $8.686/\sqrt{2N}$. Using 4000 symbols, for example, would mean that there is a 1% chance that the actual SNR differs from the estimated SNR by more than 0.25 dB.

The first requirement turns out to be much weaker than the second, so it can be ignored.

---

[2] For channel estimation and bit loading there is, strictly speaking, no need to transmit subcarriers that can never be used for data transmission (e.g., those outside the band in an FDD system), but, as we shall see in Section 9.3, some of these subcarriers may be needed for accurate calculation of the channel impulse response.

5.3 ADAPTIVE BIT LOADING: SEEKING THE "SHANNONGRI-LA" OF DATA TRANSMISSION[3]

Several algorithms for calculating the $b(n)$ have been described; the choice of the appropriate one depends mainly on whether the system is total power limited or PSD limited. The first algorithm [Hughes-Hartogs, 1987] was developed for voiceband modems. These are total power limited because the important constraint is the power delivered to the multiplexing equipment at the CO—so many milliwattts regardless of the bandwidth used.

ADSL modems, on the other hand, are PSD limited because it is necessary to limit the crosstalk induced in other pairs. VDSL modems may be either total power or PSD limited. For both constraints the $b(n)$ may be chosen to achieve any one of the following:

1. Maximum data rate at a defined error rate, margin, and coding gain: that is, at a defined γ
2. Minimum error rate at a defined data rate, margin, and coding gain: that is, maximum γ at a defined data rate
3. Maximum data rate that is an integer multiple of some $N \times$ the symbol rate[4] under the same conditions as item 1.

5.3.1 Adaptive Loading with a PSD Limitation

Maximum Data Rate at a Defined γ. The algorithm for this requirement would appear to be very simple: the $b(n)$ can be calculated from (5.6), and then R from (5.3). The single value of γ calculated from (5.5) is exact only for square constellations [i.e., for $b(n)$ even]; for $b(n)$ odd and $\geqslant 5$ the error is less than 0.2 dB and can be ignored; for $b(n) = 1$ and 3 γ should be increased by factors of 1.5 and 1.29 respectively. The continuously variable $b(n)$ must, however, be rounded to integers, while maintaining the equality of error rate on all subcarriers in order to minimize the overall error rate. This must be done by scaling the transmit levels so as to result in new SNR values given by

$$\frac{\log_2(1 + (\text{SNR}'/\gamma))}{\log_2(1 + (\text{SNR}/\gamma))} = \frac{[b(n)]}{b(n)} \tag{5.7}$$

where $[b(n)]$ is the rounded value of $b(n)$. The scaling parameters, called $g(n)$ in T1.413, are then given by

$$g(n) = \frac{2^{[b(n)]} - 1}{2^{b(n)} - 1} \approx 2^{[b(n)] - b(n)} \tag{5.8}$$

---

[3] I know, I used this one in [Bingham, 1990], but I cannot resist repeating it!
[4] T1.413 specifies that the minimum increments of data rate should be 32 kbit/s, which is 8 × the symbol rate.

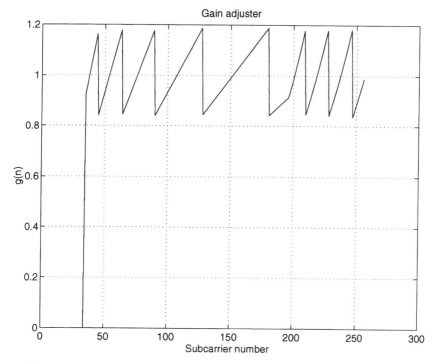

Figure 5.2   Transmit gain adjustments, $g(n)$, for a monotonically decreasing SNR.

The $g(n)$ for the simple case of monotonically decreasing SNR and $b(n)$ (from 12 on tone 30 to 4 on tone 255) are shown in Figure 5.2. The saw-toothed shape between 0.84 and 1.19 (± 1.5 dB) occurs because as the frequency increases, the rounding changes from rounding down to rounding up, and the g increases to compensate. This means that the PSD does exceed the limit in some narrow bands, but is within the limit when averaged over the range of any one value of b.

Maximum γ ***at a Defined Data Rate.***   For the second requirement it might seem that (5.6) and (5.3) could be solved to express γ as a function of B, but the nonlinear operation of rounding interferes. The following iteration is needed:

1. Calculate B_1, a first value of B from (5.3) and (5.6) using γ_1 derived from the maximum acceptable error rate.
2. If $B_1 <$ the desired B_{des}, then reporting to, and renegotiation with, the higher layers may be needed.
3. If $B_1 > B_{des}$, calculate γ_{k+1} from

$$\gamma_{k+1} = \gamma_k \times 2^{(B_k - B_{des})/Ncar_k} \qquad \text{for } k = 1, \dots$$

where $Ncar_k$ is the number of subcarriers used on the kth iteration.
4. Repeat steps 1 to 3 as needed.

Because of the rounding operation this algorithm may oscillate about the desired B. Every programmer will have his or her favorite and proprietary way of avoiding this!

Maximum Data Rate That Is an Integer Multiple of Some $N \times$ the Symbol Rate. The smallest increment of data rate for the mod/demod part of an MCM system is the symbol rate (adding one bit on one subcarrier), but the R-S FEC code words, which usually are locked to the symbol rate, are made up of bytes. Therefore, B is usually constrained to be an integer multiple of $8 \times f_s$. Then the preceding algorithm should be modified:

1a. Calculate B_0 from (5.3) and (5.6) using γ_0 derived from the maximum acceptable error rate.

1b. Truncate B_0 to $B_1 =$ the nearest multiple of $8 \times f_s$, and continue as previously.

5.3.2 Adaptive Loading with a Total Power Constraint

The basic principle of this algorithm, which is similar to, but slightly simpler than, that in [Hughes-Hartogs, 1987], is that the loading is increased one bit at a time, and each time the new bit is added to the subcarrier that requires the least additional power. This ensures that any accumulated data rate is transmitted by the minimum power.

An interesting small difference between the PSD-limited and total power-limited cases is that for the latter one-bit constellations need not be considered. The power needed for one 4QAM subcarrier is the same as for two two-point subcarriers, so it is better to use the narrower band signal.[5]

The algorithm is therefore initialized by calculating the power needed for two bits on each subcarrier. The SNR needed for two bits is 3γ. Therefore, if the noise power measured on tone k is $\mu(k)^2$, the received signal power needed for two bits is $3\gamma\mu(k)^2$, and the power that must be transmitted to deliver this is

$$
\begin{aligned}
P(1,k) &= \frac{3\gamma\mu(k)^2}{|H(k)|^2} \\
&= \frac{3\gamma P_{sc}}{\mathrm{SNR}(k)}
\end{aligned}
\tag{5.9}
$$

where P_{sc} is the power transmitted per subcarrier during channel measurement ($= 4312.5 \times 10^{-4}$ mW for ADSL). Then the incremental transmit powers needed for subsequent bits on that subcarrier can be defined by successive

---

[5] The benefit from using just one one-bit subcarrier at the edge of the band is insignificant.

TABLE 5.1 Powers and Incremental Powers for Multipoint Constellations

				m			
3	4	5	6	7	8	9	10
P 6	10	20	42	82	170	330	682
ΔP 4	4	10	22	40	88	160	352
$\alpha(m)$ 4.0	1.0	2.5	2.2	1.818	2.2	1.818	2.2

multiplications beginning with $\Delta P(1,k) = P(1,k)$:

$$\Delta P(m, k) \equiv P(m, k) - P(m - 1, k)$$
$$\equiv \alpha(m)\Delta P(m - 1, k) \tag{5.10}$$

The constellations defined in T1.413 are alternating squares and crosses with a suboptimal one used for three bits.[6] The normalized powers of these and the $\alpha(m)$ for $m = 3$ to 10 are shown in Table 5.1; thereafter the αs alternate between 2.2 and 1.818. The $\alpha(m)$ can be stored in ROM, and the incremental power updated after each bit addition by just the one multiplication shown in (5.10).

NOTE: The principle of always adding onto the "cheapest" subcarrier automatically generates the saw-toothed PSD caused by the $g(n)$ of the previous algorithm.

This algorithm can be used for all three maximum data rate/minimum error rate combinations considered above. It can also be used with a PSD constraint; the process stops on each subcarrier if the next allocation would push that subcarrier over the allowed PSD.

This algorithm is better than the one in Section 5.3.1 in that it is more versatile and is guaranteed to converge to the optimum; it is worse in that it is slower—the search over all the usable subcarriers for the smallest ΔP may have to be performed as many as 1500 times—and may be covered by the Hughes-Hartogs patent.

5.4 SCM/MCM DUALITY

Time-frequency duality was discussed in [Bello, 1964], and recently much has been made of the supposed duality of SCM and MCM systems: what SCM does in the time domain MCM does in the frequency domain, and vice versa. Examples that have been cited (and argued about) include:

---

[6] A star constellation is more efficient for three bits, but the sub-optimal one has the advantage that all the points lie on the square grid, which simplifies coding and decoding.

1. A single tone of interference ("in the frequency domain") was originally said to be the dual of an impulse of noise ("in the time domain"), but [Werner and Nguyen,1996] pointed out that it really is the dual of a repeated sequence of impulses. Without any measures to correct for it, a single tone of sufficient amplitude would wipe out a few subcarriers and cause a very high error rate; similarly, a sequence of impulses would wipe out a single-carrier system.

2. Conversely, a sequence of small impulses of noise would be spread evenly over all subcarriers, in the same way that a single interfering tone would be spread harmlessly over all time in a single-carrier system; neither would cause errors.

3. On the other hand, a sequence of large impulses would cause errors on all subcarriers in the same way that a large interfering tone would cause errors in all single-carrier symbols.

4. Ideal SCM pulses are limited in bandwidth and infinite in duration; they maintain orthogonality because they are zero at regular sampling instants. Ideal MCM pulses that are generated by an IDFT (see Chapter 6) are limited in time and infinite in bandwidth; they maintain orthogonality because they are zero at regular frequency intervals.

5. In Chapter 7 we consider a system that uses partial response in the frequency domain; this is a dual of the well-known SCM time-domain partial response systems described in [Lender, 1964] and [Kretzmer, 1965].

It must be recognized, however, that the concept of duality is useful only as a tool for early learning and perhaps later inspiration, and furthermore, only to those for whom intuition is an important part of understanding. For any particular problem (e.g., analysis of the effects of impairments) MCM must be analyzed with the same degree of rigor that has been applied to SCM problems—without invoking the fact (or, more probably, the opinion) that the problem is or is not the dual of one in SCM.

5.5 DISTORTION, EFFICIENCY, AND LATENCY

Transients occur only at the beginning and end of a multicarrier symbol, so for a channel with an impulse response of a given duration, the effects of distortion can be diluted by increasing the symbol length. This effect is quantified for the "filterless" implementation of MCM in Section 6.1. The processing time through a multicarrier transmitter and receiver is typically about five symbols.[7] The maximum length of a symbol is therefore limited to about 0.2 times the permissible latency. Other methods of reducing the effects of distortion are:

---

[7] It can be shortened by clever use of buffering and memory, but not by much.

- Use of a cyclic prefix (Section 6.2)
- Time-domain equalization (Sections 7.3.4 and 9.3)
- Sidelobe suppression (Sections 9.2, 9.3, and 9.4)

5.6 THE PEAK/AVERAGE RATIO PROBLEM

If the N subcarriers of a multicarrier signal each have unit average power and are each modulated with just 4QASK (two bits), the root-mean-square and maximum output samples are $\sqrt{N}$ and $N\sqrt{2}$ respectively; the peak/average ratio (PAR) that results is $\sqrt{2N}$. If the carriers were all modulated with multipoint QASK constellations, which themselves have a PAR that approaches $\sqrt{3}$ for large constellations, the output PAR would be $\sqrt{6N}$.

For the downstream ADSL signal and for both down and up VDSL signals, $N = 512$, which would result in a theoretical PAR $\simeq 55$; the summation of so many individual sine waves should, however, ensure that the central limit theorem applies, and the amplitude distribution of the signal for all probabilites of interest can be considered to be Gaussian. For the upstream ADSL signal, $N = 64$ and the absolute PAR $= \sqrt{384} = 19.6$. This would again seem to be large enough to ensure a Gaussian distribution over the interesting range of amplitudes,[8] but I have heard reports that the real distribution is broader than Gaussian (i.e., the tails are higher). Checking this would, however, require either the simulation or the measurement of many millions of samples of the output signal (both very tedious), and I have seen neither confirmation nor refutation of this; in the interests of simplicity we will assume a Gaussian distribution.

Such a distribution would, of course, have an infinite PAR, but the peaks would occur only once in an aeon! All MCM systems must therefore decide on some PAR and be prepared to deal with the clips that occur if the signal exceeds that. PARs values less than 3.0 are almost certainly not practically attainable; PARs greater than 7.0 are, as we shall see, expensive and unnecessary. Choosing a number in that 3 to 7 range is an important preliminary task in the design of multicarrier systems.

NOTES:

1. SCM PARs are typically calculated in the baseband and must be increased by $\sqrt{2}$ (3 dB) to account for the modulation into a passband. It is sometimes said (e.g., in [Saltzberg, 1998]) that MCM PARs must be similarly increased, but this is incorrect. It is the output samples that are Gaussian distributed; whether they are considered baseband or passband signals is irrelevant.

---

[8] Typically for probabilities $>10^{-9}$, that is, out to about the 8σ point.

2. Whether the PAR of MCM signals is higher, significantly higher, or—as claimed in some of the more partisan writings—disastrously higher than that of SCM signals is a very controversial subject. [Saltzberg, 1998] points out that the sharp bandlimiting filters (typically 15% excess bandwidth) used in SCM systems may increase the PAR calculated from simple modulation of a baseband constellation by as much as 6 dB: bringing the two systems to about the same PAR values. Cioffi[9] has suggested that this can be thought of as a Gaussian distribution induced by the multiple taps of the bandlimiting filters.

The main disadvantage of the high PAR of any signals, MCM or SCM, is that all components, analog and digital, must have a wide dynamic range. In order of increasing importance, this affects:

- *The DSP*. Providing one more bit of internal processing precision may, depending on the processor, be either inconsequential or extremely irksome.
- *The converters*. The digital-to-analog (DAC) and analog-to-digital (ADC) converters may need to accommodate one more bit than does a single-carrier *unfiltered* signal. There are, however, ways of ameliorating this problem; these are described in Section 7.2.5.
- *The quality of POTS service with some telephone handsets if no splitter is used*. This problem is discussed in more detail in Chapter 9.
- *The analog front-end circuitry*. Increasing the peak output voltage does not significantly increase either the output power or the *active power* consumed in the line drivers. The *quiescent* power consumed in the bias circuits of conventional drivers is, however, proportional to the square of this peak voltage. There are ways of reducing this bias power (many of them proprietary), but there is, nevertheless, a very strong motivation to reduce the output PAR.

Methods of reducing the PAR are discussed in Section 8.2.11; for the moment we concentrate on defining the problem.

5.6.1 Clipping

If the PAR is set at k, the probability of a clip is

$$\mathrm{Pr}_{\mathrm{clip}} = \sqrt{2/\pi} \int_k^\infty e^{-x^2/2} \tag{5.11}$$

---

[9] Private conversation.

and the average energy in a clip (assuming unit signal energy per sample[10]) is

$$E_{clip} = \sqrt{2/\pi} \int_k^{\infty} x^2 e^{-x^2/2} \qquad (5.12)$$

Early ADSL systems used a PAR of about 6.0 (15.6 dB); 4.0 (12 dB) is probably about the maximum that will be acceptable for a second-generation system, 3.5 (10.9 dB) is a reasonable number to strive for, and 3.0 (9.5 dB) is an aggressive goal for a G.lite system. We will therefore consider k values over a range 3.0 to 6.0.

For these values of k, E_{clip} is very small and the signal/clip energy ratio (SCR) is high. It might seem, therefore, that the contribution of clip energy to the total noise would be insignificant. This is, however, very misleading; the clip energy must not be averaged over all symbols, but only over those symbols in which a clip occurs. The probability of a clip occurring in a symbol is

$$Pr_{clipsymb} = 1 - (1 - Pr_{clip})^N \qquad (5.13)$$

and since the signal energy per sample is normalized to unity, the signal/conditional clip ratio is

$$SCCR = \frac{Pr_{clipsymb}}{E_{clip}} \qquad (5.14)$$

$Pr_{clipsymb}$ and SCCR are shown in Table 5.2 for values of k from 3.0 to 6.0 for the case $N = 512$.

In every symbol in which any number of clips occur, noise will be spread evenly across the entire band and will eventually cause errors on all subcarriers for which the loading is such that $[3b(n) - 3] > SCCR$. As an example, if the maximum loading is 12 bits, errors will occur in all symbols for which SCCR < 33 dB. It must be noted, however, that the SCCRs in Table 5.2 are average values; the total clip energy and the resultant SCCR will vary widely from one

TABLE 5.2   Clip Probabilities and SCCR Values for Various PARs

	PAR (k)						
	3.0	3.5	4.0	4.5	5.0	5.5	6.0
Pr(clipsymb)	0.75	0.21	3.2×10^{-2}	3.5×10^{-3}	2.9×10^{-4}	1.9×10^{-5}	10^{-6}
Average SCCR (dB)	32.7	35.8	37.1	38.0	38.8	39.5	40.2

---

[10] Normalizing with respect to the signal energy rather than power is simpler because it makes the explanation independent of the sample rate.

clipped symbol to the next. The statistical distribution of the SCCRs is difficult to calculate, but preliminary calculations show that the 1 percentile, for example, may be 12 dB or more below the average. This suggests that it would be almost impossible to prevent errors *when clips occur* and that the only feasible strategy is to set the PAR high enough to make the clip rate acceptably low. Most first-generation ADSL transmitters played it safe and used PARs of 6.0 (15.6 dB) or more, but the methods of PAR reduction and clip shaping discussed in Section 8.2.11 should improve matters greatly for second-generation ADSL and VDSL.

Entropy of a Clip. Clips are very tantalizing because as discussed in more detail in Chapter 7 of [Starr et al., 1999] and many other papers referenced therein, the loss of capacity that would be incurred by detecting a clip in the transmitter and conveying all information about it to the receiver is very small. Even with a PAR of 10 dB, the *theoretical* loss is only about 3%.

6

DFT-BASED MCM
(MQASK, OFDM, DMT)

[Weinstein and Ebert, 1971] described the simplest way of performing the modulation shown generally in Figure 5.1. There is no filtering of the output of the constellation encoders, and the real and imaginary parts of each of the N_{car} words are used to quadrature amplitude shift key (QASK) the N_{car} tones. An example of the QASKing of tone 8 with three successive symbols is shown in Figure 6.1.

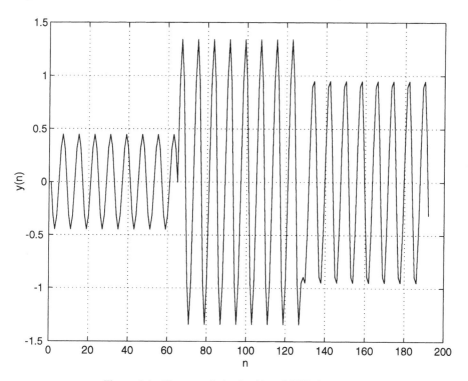

Figure 6.1 Three symbols of a 16-pt QASKed tone 8.

The multiple QASK operation is performed using an inverse discrete Fourier transform (IDFT). Important things to note about this transform are:

- The input comprises $(N_{car} - 1)$ complex numbers, which are quadrature modulated onto tones 1 to $(N_{car} - 1)$, plus two real numbers, which are modulated onto dc and tone N_{car}.
- It generates $N = 2N_{car}$ real samples and is called an N-point IDFT.
- It can be performed efficiently as an IF(ast)FT,[1] as described in Appendix C.

Subsequent implementations of MQASK, with different names, were described in [Keasler and Bitzer, 1980], [Hirosaki, 1981], and [Fegreus, 1986], but the only two that have survived are orthogonal FDM (OFDM) for wireless use (see the specialized bibliography at the end of the references) and DMT for DSL. I will use the name MQASK whenever the emphasis is on the rectangular nature of the envelope, and DMT when discussing overall systems.

The baseband pulse is a rectangle of duration T, and the DFT of keyed tone n, $F(n,k)$, comprises terms with amplitudes proportional to $\text{sinc}(n-k)$ and $\text{sinc}(n+k)$, but the phases of these and how they combine depends on the phase of the keyed sinusoid. We are mainly interested in the PSD of the signal, so it is convenient to write

$$|F(n,k)|^2 = \text{sinc}^2(n-k) + \text{sinc}^2(n+k) \qquad (6.1)$$

The $(n+k)$ term causes a slight asymmetry about $k = n$, but for the values of n used for xDSL (typically $\geqslant 7$), it is insignificant. Figure 6.2 shows $|F(n,k)|^2$ for $n = 8$ with k treated as a continuous variable. It also shows the powers averaged across each subchannel band centered at integer values of k and width $\Delta f\,(= 1/T)$. That is,

$$|F_{\text{smothed}}(n,k)|^2 = \int_{m=-0.5}^{m=+0.5} |F(n,k+m)|^2\, dm \qquad (6.2)$$

These are more indicative of the sidelobe magnitudes than the alternating zeros and peaks. As would be expected, the sum of the left-hand sides over all k values is unity. It will be useful later to have a simple mathematical model for this smoothed spectrum:

$$|F_{\text{smoothed}}(n,k)|^2 \approx \frac{1}{2\pi^2(|k - n| - 1)} \qquad \text{for} |k - n| \geqslant 2 \qquad (6.3)$$

This is also shown in Figure 6.2 as superimposed '×'s.

---

[1] Strictly speaking, the D refers to the algorithm and the F to the implementation, but for simplicity we will use the F for both from now on.

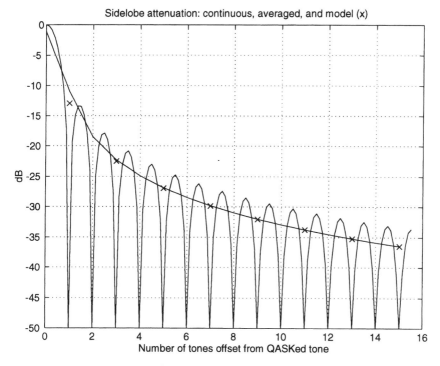

Figure 6.2   Sidelobe attenuation of QASKed tone 8.

Receiver. The matched receiver for a QASK signal is a demodulator followed by a complex integrate and dump; these operations can be performed by an FFT.

6.1   GUARD PERIOD

If an MQASK signal is passed through a channel with a finite IR, the envelope of every tone—each ideally rectangular—will be different. It is, however, useful to consider a generic envelope as shown in Figure 6.3; this suggests two ways of using an extra ν samples to overcome the distortion caused by the finite IR.

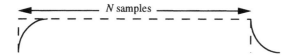

Figure 6.3   Distorted QASK envelope.

1. Add the last ν samples to the beginning of the signal (a *cyclic prefix*), and then delay collecting the N samples in the receiver until the transient response has finished.

2. Add the first ν samples to the end of the signal (a *cyclic suffix*), and then similarly delay collecting the samples. The only difference between this and method 1 is that the apparent phase of every tone is delayed by $2\pi n\nu/N$.

The most important property of all envelopes—that the trailing edge is the complement of the leading edge—suggests a third way of overcoming distortion.

3. Use a quiet period between symbols, and in the receiver add the last ν samples to the first.

All three methods result in zero sensitivity to IR terms h_i with $i \leqslant \nu$, and the same sensitivity to terms with $i > \nu$ (see Section 6.2). The important differences are in the spectra of the transmitted signals and the effective transfer functions of the receivers:

1,2. *Prefix and suffix.* The frequency separation of the zeros of the output spectrum is $N/T(N+\nu)$. Since the tone separation is $1/T$, this means that the spectral zeros do not coincide with the tone frequencies.[2] The frequency separation of the zeros of the receiver transfer function, on the other hand, is $1/T$, so these zeros do coincide with the tone frequencies.

3. *Quiet period.* Because the transmitted pulse is of duration T, the frequency separation of the zeros of the output spectrum is $1/T$. Conversely, because the receiver uses all $(N + \nu)$ samples, the frequency separation of the zeros of the receiver transfer function is $N/T(N+\nu)$.

These frequency separations become significant when methods of shaping the output PSD (see Section 6.5) and canceling RFI (Sections 10.6.4 and 11.6) are considered.

The guard period wastes ν samples, so the data rate efficiency is

$$\varepsilon = \frac{1}{1 + \nu/N} \tag{6.4}$$

How this translates into decibels of margin depends on the number of bits/s/Hz. For ADSL $N = 512$ and $\nu = 32$, so $\varepsilon \approx 0.94$. Therefore, in round numbers:

---

[2] The orthogonality of the MQASK signals is sometimes "explained" by saying that the spectrum of each is zero at that other tone frequencies. This is clearly not the reason, because with a cyclic prefix the zeros do not fall at the tone frequencies, yet orthogonality is achieved!

- For a data rate of 6 + Mbit/s using 1 MHz of bandwidth 0.35 bit/s per Hertz are wasted, for a loss in margin of approximately 1 dB.
- For a data rate of 1.5 + Mbit/s on a long loop using 0.5 MHz of bandwidth, the loss would be approximately 0.5 dB.

6.1.1 Length of the Guard Period

In some of the early writings on MCM it was argued that if the duration of the guard period exceeds the variation of the group delay across the band, all keyed tones will have "arrived" at the receiver by the end of the guard period, and orthogonality will be ensured. A simple counterexample to this is a channel that has a single pole; the group delay will have some finite maximum, but the IR will have infinite duration, and orthogonality will not be preserved. Choice of the length of the guard-period requires a compromise between efficiency as defined by (6.4) and ease of designing the equalizer (see Sections 8.4.4 and 11.2).

6.2 EFFECTS OF CHANNEL DISTORTION

Because no practical channels have a finite IR, the ideal of an IR that is contained within the cyclic prefix is never achieved in practice. The following analysis of the effects of a longer IR was first described in [Jacobsen, 1996].

For all xDSL systems it is an acceptable approximation to limit the length of the theoretically infinite IR to the symbol length (N samples). Therefore, let the causal IR of the channel[3] be defined as h_i for $0 \leqslant i \leqslant N - 1$, and let a cyclic prefix of ν terms be used. The equalization and timing recovery process described in Section 8.4.4 selects the block of $(\nu + 1)$ contiguous h terms that contains the maximum energy. That is, it finds the value of k for which the "windowed" energy ($h_k^2 + h_{k+1}^2 + \cdots + h_{k+\nu}^2$) is maximized. In general, the best value of k will not be zero (i.e., there will be both pre- and postcursors), but for the sake of clarity in this first explanation we will assume that it is.

Let us also use a simple, specific set of parameters for the DMT system to be analyzed[4]: $N = 8$ and $\nu = 3$ for a total symbol length of 11. If the mth symbol set before the cyclic prefix is added is defined as $[x_{m,i}]$ for $i = 1$ to 8, then after transmission through the channel the samples of interest for the reception of the $(m + 1)$th symbol are

$$y = [x_{m,6}, x_{m,7}, x_{m,8}, x_{m,1}, \ldots, x_{m,8}, x_{m+1,6}, x_{m+1,7}, x_{m+1,8}, x_{m+1,1}, \ldots, x_{m+1,8}]$$
$$* [h_0, h_1, \ldots, h_7] \qquad\qquad (6.5)$$

---

[3] Including both converters, all filters, and the loop.
[4] Extrapolation to the general, or any other specific, case should be easy.

After stripping off the (now distorted) cyclic prefix, the column vector $[y_{m+1}]$ for input to the DFT can be written as the sum of two vectors:

$$
\begin{bmatrix} y_{m+1,1} \\ y_{m+1,2} \\ y_{m+1,3} \\ y_{m+1,4} \\ y_{m+1,5} \\ y_{m+1,6} \\ y_{m+1,7} \\ y_{m+1,8} \end{bmatrix} = \begin{bmatrix} h_0 & 0 & 0 & 0 & 0 & h_3 & h_2 & h_1 \\ h_1 & h_0 & 0 & 0 & 0 & h_4 & h_3 & h_2 \\ h_2 & h_1 & h_0 & 0 & 0 & h_5 & h_4 & h_3 \\ h_3 & h_2 & h_1 & h_0 & 0 & h_6 & h_5 & h_4 \\ h_4 & h_3 & h_2 & h_1 & h_0 & h_7 & h_6 & h_5 \\ h_5 & h_4 & h_3 & h_2 & h_1 & h_0 & h_7 & h_6 \\ h_6 & h_5 & h_4 & h_3 & h_2 & h_1 & h_0 & h_7 \\ h_7 & h_6 & h_5 & h_4 & h_3 & h_2 & h_1 & h_0 \end{bmatrix} \begin{bmatrix} x_{m+1,1} \\ x_{m+1,2} \\ x_{m+1,3} \\ x_{m+1,4} \\ x_{m+1,5} \\ x_{m+1,6} \\ x_{m+1,7} \\ x_{m+1,8} \end{bmatrix}
$$
$$
+ \begin{bmatrix} 0 & 0 & 0 & 0 & h_7 & h_6 & h_5 & h_4 \\ 0 & 0 & 0 & 0 & 0 & h_7 & h_6 & h_5 \\ 0 & 0 & 0 & 0 & 0 & 0 & h_7 & h_6 \\ 0 & 0 & 0 & 0 & 0 & 0 & 0 & h_7 \\ 0 & 0 & 0 & 0 & 0 & 0 & 0 & 0 \\ 0 & 0 & 0 & 0 & 0 & 0 & 0 & 0 \\ 0 & 0 & 0 & 0 & 0 & 0 & 0 & 0 \\ 0 & 0 & 0 & 0 & 0 & 0 & 0 & 0 \end{bmatrix} \begin{bmatrix} x_{m,1} \\ x_{m,2} \\ x_{m,3} \\ x_{m,4} \\ x_{m,5} \\ x_{m,6} \\ x_{m,7} \\ x_{m,8} \end{bmatrix}
\tag{6.6}
$$

The second vector clearly represents intersymbol interference (the effect of x_m on y_{m+1}), but for easiest understanding of the detection method, the first vector, which represents the effects of the "present" symbol, should be split into two:

$$
y_{m+1} = \mathbf{H}x_{m+1} + \mathbf{H}_0 x_{m+1} + \mathbf{H}_1 x_m
\tag{6.7}
$$

where

$$
\mathbf{H} = \begin{bmatrix} h_0 & h_7 & h_6 & h_5 & h_4 & h_3 & h_2 & h_1 \\ h_1 & h_0 & h_7 & h_6 & h_5 & h_4 & h_3 & h_2 \\ h_2 & h_1 & h_0 & h_7 & h_6 & h_5 & h_4 & h_3 \\ h_3 & h_2 & h_1 & h_0 & h_7 & h_6 & h_5 & h_4 \\ h_4 & h_3 & h_2 & h_1 & h_0 & h_7 & h_6 & h_5 \\ h_5 & h_4 & h_3 & h_2 & h_1 & h_0 & h_7 & h_6 \\ h_6 & h_5 & h_4 & h_3 & h_2 & h_1 & h_0 & h_7 \\ h_7 & h_6 & h_5 & h_4 & h_3 & h_2 & h_1 & h_0 \end{bmatrix}
\tag{6.8}
$$

$$
\mathbf{H}_0 = - \begin{bmatrix}
0 & h_7 & h_6 & h_5 & h_4 & 0 & 0 & 0 \\
0 & 0 & h_7 & h_6 & h_5 & 0 & 0 & 0 \\
0 & 0 & 0 & h_7 & h_6 & 0 & 0 & 0 \\
0 & 0 & 0 & 0 & h_7 & 0 & 0 & 0 \\
0 & 0 & 0 & 0 & 0 & 0 & 0 & 0 \\
0 & 0 & 0 & 0 & 0 & 0 & 0 & 0 \\
0 & 0 & 0 & 0 & 0 & 0 & 0 & 0 \\
0 & 0 & 0 & 0 & 0 & 0 & 0 & 0
\end{bmatrix}
\tag{6.9}
$$

H is a circulant matrix, and the first term of (6.7) would be the only one if the input x were cyclic *or* if the cyclic prefix were long enough to span the IR. The subscripted **H**'s represent distortion: $\mathbf{H}_0$ defines the interference of the "present" symbol with itself (i.e., intrasymbol/interchannel interference), and $\mathbf{H}_1$ defines the interference from the "previous" symbol (i.e., intersymbol/interchannel interference). It can be seen that each contains h_i only with $i > v$; that is, no distortion results from IR terms within the range of the cyclic prefix. Also $\mathbf{H}_0$ and $\mathbf{H}_1$ contain the same distribution of h_i terms, albeit in different places.

6.2.1   Total Distortion: Signal/Total Distortion Ratio

If the average energy of each transmit sample, $x_{m,i}$ is normalized to unity, the total signal energy of the $(m + 1)$th symbol is

$$
|H|^2 = N \sum_{i=0}^{7} h_i^2
\tag{6.10}
$$

and the total distortion energy— contributed equally by the mth and $(m +1)$th symbols—can be seen from (6.9) and (6.6), respectively, to be

$$
|H_0|^2 + |H_1|^2 = 2(h_4^2 + 2h_5^2 + 3h_6^2 + 4h_7^2)
\tag{6.11}
$$

Therefore, since $N = 2N_{\mathrm{car}}$, the signal/total distortion ratio (STDR)[5,6] is

$$
\mathrm{STDR} = N_{\mathrm{car}} \sum_{i=0}^{7} h_i^2 \Big/ \sum_{i=4}^{7} (i - 3)h_i^2
\tag{6.12}
$$

[5] The STDR is a single wideband measure of distortion; in Section 6.2.3 we consider the SDRs on the individual subchannels.

[6] The signal and distortion can be considered as passing through the "window" and being splattered on the "wall," respectively.

which can be generalized to

$$STDR = N_{car} \sum_{i=0}^{N} h_i^2 \Bigg/ \sum_{i=\nu+1}^{N} (i-\nu)h_i^2 \qquad (6.13)$$

If we wish to compare this STDR to that for a single-carrier system ($N_{car} = 1$), we should set the length of the DFE equal to ν, so that both systems will be immune to IRs shorter than ($\nu + 1$). Then

$$STDR_{MCM,\nu=0} = N_{car} \sum_{i=0}^{N} h_i^2 \Bigg/ \sum_{i=\nu+1}^{N} (i-\nu)h_i^2 \qquad (6.14)$$

but

$$STDR_{SCM} = h_0^2 \Bigg/ \sum_{i=\nu+1}^{N} h_i^2 \qquad (6.15)$$

For N large, $STDR_{MCM} \gg STDR_{SCM}$. This *should* make the equalization task much easier, but as we shall see, there are many factors that must be taken into account. It is interesting (and perhaps counterintuitive) that the performance of an MCM system, in which all the terms of the IR contribute to the signal energy, is only as good as that of an DF-equalized SCM system, in which only the first term contributes!

6.2.2 Case of Both Post- and Precursors

Having established the ramp weighting of the energy for an IR with only postcursors, it is easy to generalize to the case where there are IR terms both before and after the selected window. If, in our previous example, the set $[h_2, \ldots, h_5]^7$ were chosen, the total distortion would be given by

$$|wall|^2 = 2.(2h_0^2 + h_1^2 + h_6^2 + 2h_7^2) \qquad (6.16)$$

and the specific- and general-case denominators of (6.12) and (6.13) would be changed appropriately. This ramp weighting of the walls is shown informally in Figure 6.4.

6.2.3 Distortion on Individual Subchannels: SDR(*j*)

The wideband STDR is not, however, the complete measure of the effects of distortion in an MCM system, because the total distortion will be distributed unequally among the subchannels, and the effects of that distributed distortion

---

[7] Being careful not to use negative subscripts for *h*, which would imply that the IR was noncausal.

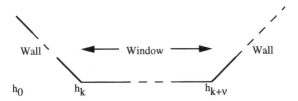

Figure 6.4   Weighting of window and walls.

will depend on the SNRs of the subchannels. For example, an SDR of 20 dB on a subchannel that carries only 2 bits would be fairly inconsequential; on a subchannel carrying 12 bits it would be disastrous.

Calculation of SDR(f) is a tedious process; it involves:

1. Convolving the QASK signal (including the cyclic prefix) for every used tone with the IR (equalized if appropriate) of the channel
2. Discarding the cyclic prefix portion and FFTing the remaining N samples to generate the response to each tone and the *contribution* of that keyed tone to the total distortion
3. Accumulating the contributions for all used tones

Undoubtedly, several DMT system designers have done this, but detailed results have remained proprietary. A reportable result is that for the ADSL upstream channel, the distortion resulting from the loop, the transformers, and the POTS-protecting high-pass filters is such that at the low end of the band (typically, tones 7 to 11, as defined in Section 8.1.2), SDR < SNR, and there is a significant loss of capacity.

Unfinished Business. The long path from the parameters of an equalizer to the effects on the overall channel capacity makes optimizing the design of an equalizer nearly impossible. This path might be shortened and/or smoothed, or a frequency-domain equalizer (see Section 11.2.2) might be used to link cause and effect more closely.

6.3   THE SIDELOBE PROBLEM

As can be seen from Figure 6.2, the sidelobes of an MQASK modulator or demodulator fall off slowly. This has several consequences, which we consider in turn.

6.3.1   Noise Smearing and Resultant Enhancement

The windowing process in the receiver spreads the input noise in any one subchannel over many. Consider two subchannels, m_1 and m_2. Some of the noise in m_1 will appear in the DFT output of m_2, and vice versa. If the noise in m_1 is

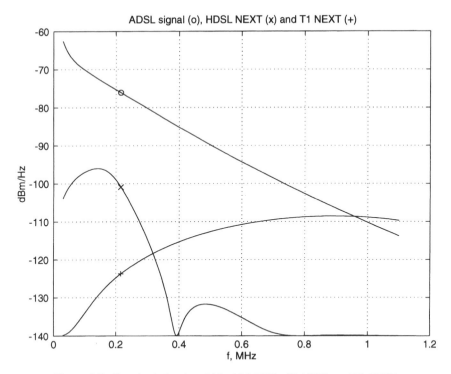

Figure 6.5   Received signal on 9 kft of 26 AWG with HDSL and T1 NEXT.

much higher than that in m_2, the sidelobes of the higher noise may contribute significantly to the lower noise, and in extreme cases, even become the dominant contributor. If the input SNR in m_1 and m_2 is the same, this noise smearing will have little effect on the total capacity, but if the SNR in m_2 is higher than that in m_1, m_2 will lose more in capacity than m_1 will gain.

Figure 6.5 shows the level of the received signal on 9 kft of 26 AWG. It also shows the NEXT from 20 HDSL interferers (one of the test situations defined in T1.413) and from 4 T1 interferers in an adjacent binder-group (very severe interference on that length of loop). Figure 6.6 shows the SNRs for HDSL NEXT at the input to the receiver and at the output of the DFT when all the noise powers have been smeared by the $F_{smoothed}$ defined in (6.3); it can be seen that the noise from about 380 kHz upward (the region where the SNR is highest) has been increased by an average of about 5 dB. The noise-smearing effect would be moderately serious.

In practice, the front end of the receiver may contain an equalizer of some sort, which would amplify the signal at higher frequencies. Since it will also amplify the noise there, it will eliminate the harmful effects of smearing the low-frequency noise into the higher frequencies. In most cases, however, smearing of high-frequency noise into the lower frequencies is, as we shall see in Section 6.3.2, much more serious.

THE SIDELOBE PROBLEM   **101**

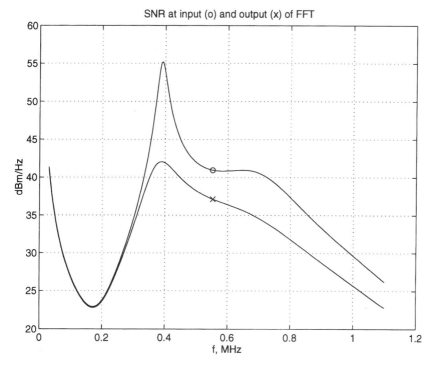

Figure 6.6   SNRs with HOSL NEXT: without equalization.

6.3.2   Noise Enhancement from Linear Equalization

The most basic linear equalizer for any modem receiver is a linear two-port that equalizes the amplitude and phase of the channel over some used band so that the impulse response of the tandem connection of channel and two-port is an impulse. The zero-forcing equalizer has been discarded for many high-performance single-carrier systems because of its noise enhancement. A method that minimizes the sum of the residual distortion and noise can do a better job, but for channels with severe amplitude distortion the loss of capacity due to noise enhancement is still typically too great to be tolerable. The decision-feedback equalizer (DFE) is a much better solution, and is well established and understood (see, e.g., [Honig and Messerschmitt,1984]).

For MCM the situation is very different, and the following argument has often been used to try to show that a linear equalizer would be adequate.

- The capacity of the full channel is the sum of the capacities of the subchannels.
- The capacity of each subchannel depends only on the SNR of that subchannel.

- This SNR is not changed by linear equalization because signal and noise are amplified or attenuated equally.
- Therefore, the capacities of all the subchannels and of the full channel are not changed.

This argument is wrong because, although the equalization itself is linear, the overall detection process is not. In Section 6.3.1 we saw that because of the slow decay of the sidelobes of a rectangularly windowed DFT, colored input noise may result in noise amplification at some frequencies (often, those that have the highest input SNR), and loss of capacity.

If the coloring of the noise is due only to crosstalk transfer functions, the effect is, as we saw in Section 6.3.1, only moderate, but if already colored noise is further colored by equalization the effect may be serious. Figure 6.5 shows the received ADSL downstream signal and noise levels before equalization, for an extreme example of 9 kft of 26 AWG with 4 T1s in an adjacent binder[8]; Figure 6.7 shows the SNRs after equalization at both the input and output of the DFT. It can be seen that the smearing would greatly reduce the SNR at low frequencies.

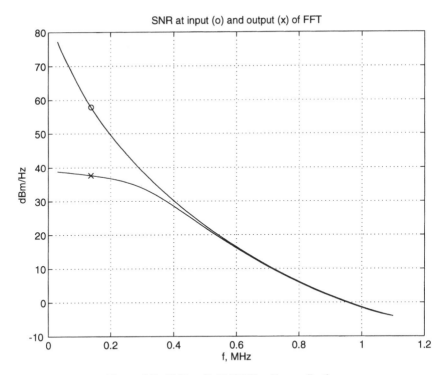

Figure 6.7   SNRs with TI NEXT: with equalization.

[8] Signal and noise are monotonically decreasing and increasing, respectively, with frequency, which results in the greatest coloring of the noise input to the FFT.

Depending on whether FDD or EC were used, this would probably result in a loss of 20 to 30% of capacity.

A crude predictor of the amount of noise enhancement is the sum of the equalization and the noise coloration (both in decibels) across the full band. A simple requirement would be that the extra noise that is spread into the lowest bin should be less than the noise that is already there; that is, it less than doubles the noise there and reduces the capacity by less than one bit. To satisfy this across a band of approximately 250 subcarriers

$$\Delta dB_{noise} + \Delta dB_{eq} < 45\,dB \qquad (6.17)$$

where ΔdB_{noise} and ΔdB_{eq} are the increases in noise PSD and equalizer gain from one end of the band to the other.

NOTE: This calculation must be performed across the full band even if the upper subbands are unusable because of low SNR. Unless filtered out before the FFT, the noise up there will still be spread into the lower subbands.

One of the advantages of a guard period (quiet period or cyclic prefix) is, as we shall see in Section 8.4.3, that it is not necessary to completely equalize an input signal. An algorithm for designing a "partial" equalizer should strive to minimize this noise enhancement effect.

6.3.3   Reducing Noise Enhancement

Noise enhancement can be reduced in three ways:

1. *Increasing the size of the DFT* (i.e., reducing Δf). Doubling the size of the DFT (halving Δf) would double the sidelobe number at any frequency and thereby reduce the magnitude of all sidelobes by approximately 6 dB. Crude calculations suggest, however, that for a typical ADSL system with as much as 50 dB variation of attenuation across the band a DFT size of at least 4096 would be needed to bring the loss of capacity in the lower subbands due to noise enhancement down to an acceptable level. This would be impractical from both memory size and latency considerations.

2. *Using a guard period.* The guard period was first used commercially in Telebit's Trailblazer voiceband modem. The FFT size was 2048, and the attenuation distortion across a voiceband is usually less than 15 dB. Noise enhancement was not serious, and the only purpose of the guard period was to simplify the equalizer; shortening the impulse response to $(\nu + 1)$ samples obviously required fewer taps than shortening it to 1. For DSL the situation is different. ADSL latency requirements limit the symbol duration to about 250 μs, and therefore the DFT size to 512; as we have

seen, equalization to an impulse (with a linear equalizer of infinite complexity!) would seriously degrade performance. A guard period allows for a very wide choice of shortened impulse responses[9] (SIRs), and in doing so, greatly simplifies the design of the equalizer. Just as important, it may also reduce the noise enhancement to a tolerable level. The characteristics of the SIR and the design of an equalizer to produce this are discussed in Section 11.2.1, but a little pre-motivation may be useful. Figure 6.8 shows the amplitude responses of the equalizers for a 9-kft 26-AWG loop for two SIRs: a 1-sample and a simple 6-sample $[(1 + 0.8D)(1 + 1.4D + 0.7D^2)(1 + 0.6D^2)]$. It can be seen that the amplitude variation across the passband—and thence the potential for noise enhancement—can be reduced significantly by even a short SIR.

3. *Using a demodulation method that attenuates the sidelobes much more rapidly.* Three such methods—SMCM, frequency-domain partial response, and DWMT—are discussed in Chapter 7, but very little has been published about the equalization of such signals.

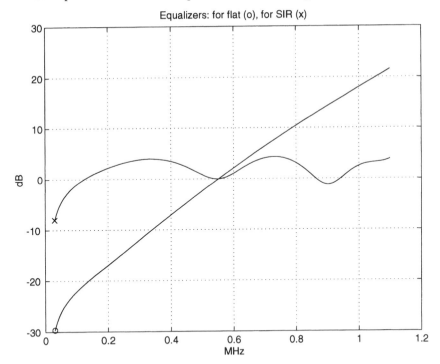

Figure 6.8 Response of equalizers of 9 kft of 26 AWG to generate two SIRs.

---

[9] The term *desired impulse response* was used in much of the literature on DFEs (e.g., [Honig and Messershmitt, 1984]), but it is misleading because it implies that the SIR can be defined a priori, which is rarely the case.

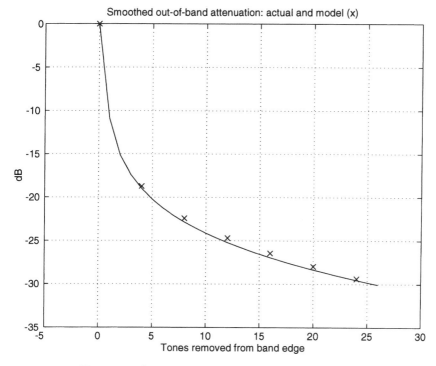

Figure 6.9 Sidelobes at the edge of an MQASK passband.

6.3.4 Band Limiting

Figure 6.9 shows the cumulative effect in the stopband of the sidelobes of a set of MQASK signals in a passband. A simple model for this that will be useful when designing filters (see Section 8.3.4) is

$$|F_{\text{stopband}}(k)|^2 = \sum_{m=k}^{N} \frac{1}{2\pi^2 m^2} \qquad (6.18)$$

where N is the total number of modulated tones in the passband, k is the number of tones removed from the band edge, and $|F_{\text{passband}}|^2$ is normalized to unity. It can be seen that turning off tones provides only a mild band limiting of an MQASK signal that will usually have to be augmented by a filter (see Section 8.3.4).

6.4 REDUCING THE SIDELOBES: SHAPED CYCLIC PREFIX

Many ways of reducing the sidelobes have been proposed; in this section we describe some simple modifications of DMT (see also [Weinstein and Ebert,1971], [Bingham, 1995], and [Spruyt et al., 1996]); Chapter 7 describes some different MCM systems. A shaped cyclic prefix is between the two extremes of unshaped cyclic prefix and a guard period discussed in Section 6.1. The ν samples of the cyclic prefix are weighted by $w(i)$, $i = 1$ to ν, and the last ν samples of the pulse are weighted by a complementary $[1 - w(i)]$. Typically, the pulse is symmetrical; that is, $[1 - w(i)] = w(\nu + 1 - i)$, and the most common shape is sine-squared (also known as raised cosine); that is, $w(i) = \sin^2(\pi i/2\nu)$.

The shaping can be done in many different ways, which differ in whether they provide the extra sidelobe attenuation in the transmit PSD or the receive transfer function, and also in the separation of the spectral zeros [Δf or $N \Delta f/(N + \nu)$]:

1. A cyclic prefix is added and shaped as described above. Figure 6.10 shows one side of the transmit PSD with $N = 512$ and $\nu = 32$: with a rectangular shaping as defined for ADSL in T1.413, and sine-squared shaping. As would be expected, shaping only 64 of the 544 samples has very little effect on the close-in sidelobes, but there is a very useful extra attenuation of the far-out sidelobes. The zeros of the PSD are separated by Δf (i.e., they fall on the tone frequencies). In the receiver, the first ν samples, instead of being discarded as in the unshaped case, are added to the last ν samples before input to the DFT. The receive transfer function has the large sidelobes, and its zeros are separated by $N\Delta f/(N + \nu)$.

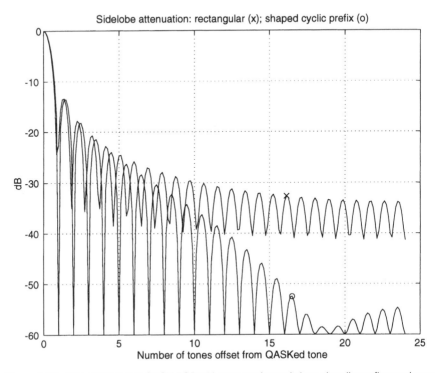

Figure 6.10   One sideband of QASK PSD with rectangular and shaped cyclic-prefix envelopes.

2. A cyclic prefix is added and the pulse envelope is transmitted unshaped. The transmit PSD has the large sidelobes, and its zeros are separated by $N\Delta f/(N+\nu)$. The shaping is applied in the receiver, and then, as in method 1, the first ν samples are added to the last before input to the DFT. The receive transfer function has the attenuated sidelobes, and its zeros are separated by Δf.

3. As in method 1, a cyclic prefix is added with its ν samples weighted by $w(i)$, and a cyclic suffix is also added, with its samples $[i=(N+\nu+1)$ to $(N+2\nu)]$ weighted by $[1-w(i)]$. The transmit PSD has the attenuated sidelobes, and its zeros are separated by $N\Delta f/(N+\nu)$. Because the envelope pulse has been extended to $(N+2\nu)$ samples, successive symbols overlap. That should not matter, however, because in the receiver all the shaped samples are discarded. The transfer function has the large sidelobes, and its zeros are separated by Δf.

4. As a combination of methods 1 and 2, half-shaping [i.e., $w(i)=\sin(\pi i/2\nu)$] can be applied in both transmitter and receiver.

6.4.1 Sensitivity to Channel Distortion

The sensitivity to the h_i terms is the same for all four methods, so we will analyze only method 1. [Jacobsen, 1996] showed that for our example of $N=8$, $\nu=3$, $\mathbf{H}$ is as given in (6.8), but if the shaping is symmetrtical, $\mathbf{H}_0$ and $\mathbf{H}_1$ as given in (6.9) and (6.6) must be modified to

$$\mathbf{H}_0 = - \begin{bmatrix} 0 & h_7 & h_6 & h_5 & h_4 & w_3h_3 & w_2h_2 & w_1h_1 \\ 0 & 0 & h_7 & h_6 & h_5 & w_3h_4 & w_2h_3 & w_1h_2 \\ 0 & 0 & 0 & h_7 & h_6 & w_3h_5 & w_2h_4 & w_1h_3 \\ 0 & 0 & 0 & 0 & h_7 & w_3h_6 & w_2h_5 & w_1h_4 \\ 0 & 0 & 0 & 0 & 0 & w_3h_7 & w_2h_6 & w_1h_5 \\ 0 & 0 & 0 & 0 & 0 & 0 & w_2h_7 & w_1h_6 \\ 0 & 0 & 0 & 0 & 0 & 0 & 0 & w_1h_7 \\ 0 & 0 & 0 & 0 & 0 & 0 & 0 & 0 \end{bmatrix} \tag{6.19}$$

$$\mathbf{H}_1 = \begin{bmatrix} 0 & 0 & 0 & 0 & h_7 & w_3h_6 & w_2h_5 & w_1h_4 \\ 0 & 0 & 0 & 0 & 0 & w_3h_7 & w_2h_6 & w_1h_5 \\ 0 & 0 & 0 & 0 & 0 & 0 & w_2h_7 & w_1h_6 \\ 0 & 0 & 0 & 0 & 0 & 0 & 0 & w_1h_7 \\ 0 & 0 & 0 & 0 & 0 & 0 & 0 & 0 \\ 0 & h_7 & h_6 & h_5 & h_4 & w_3h_3 & w_2h_2 & w_1h_1 \\ 0 & 0 & h_7 & h_6 & h_5 & w_3h_4 & w_2h_3 & w_1h_2 \\ 0 & 0 & 0 & h_7 & h_6 & w_3h_5 & w_2h_4 & w_1h_3 \end{bmatrix} \tag{6.20}$$

As in the unshaped case, the total distortion energy is contributed equally by the mth and $(m+1)$th symbols, and

$$|H_0|^2 + |H_1|^2 = 2[w_1^2 h_1^2 + (w_1^2 + w_2^2)h_2^2 + Sh_3^2 + (1+S)h_4^2 + (2+S)h_5^2 \cdots]$$

(6.21)

where

$$S = (w_1^2 + w_2^2 + w_3^2)$$

This can be generalized to

$$|H_0|^2 + |H_1|^2 = 2[w_1^2 h_1^2 + (w_1^2 + w_2^2)h_2^2 + \cdots + Sh_\nu^2 + (1+S)h_{\nu+1}^2$$
$$+ (2+S)h_{\nu+2}^2 \cdots]$$

(6.22a)

$$\equiv \sum c_i h_i^2$$

(6.22b)

where if sine-squared shaping is used (see, e.g., [Dwight, 1961]),

$$S = \sum_1^\nu w_i^2 = \frac{3\nu}{8}$$

(6.23)

The three shaping methods can be compared with unshaped DMT by plotting the c_i coefficients as shown in Figure 6.11 for the ADSL case of $N = 512$ and $\nu = 32$. It can be seen that shaping significantly increases the distortion due to IR terms from about $i = 10$ onward.

6.4.2 Advantages and Disadvantages of the Four Methods of Using a Shaped Cyclic Prefix

In summary:

- All the methods maintain orthogonality between subchannels if the channel is undistorted.
- All the methods attenuate the sidelobes of the end-to-end transfer function and thereby reduce the noise-enhancement effect in an equalizer.
- All the methods increase the sensitivity to channel distortion, and thereby, presumably, make the equalizer's task harder.
- Methods 1 and 3 are useful if the out-of-band power must be reduced without filtering; method 1 is more useful because it can be used with a dummy tone for further reduction (see Section 6.5).
- Method 2 provides the best filtering to RFI (see Section 10.6.4).
- Method 2 is the only one that is completely compatible with T1.413; it is implemented wholly in the receiver.

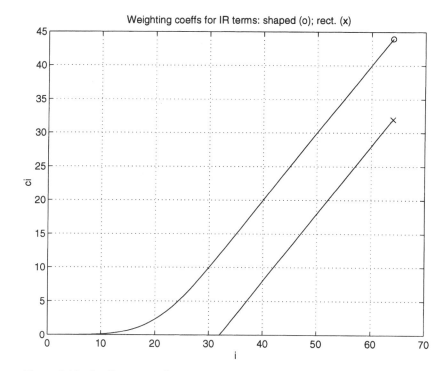

Figure 6.11 Coefficients of h_i^2 in total distortion: unshaped and shaped cyclic prefixes.

- Method 3 is interesting in that it is not, strictly speaking, standard compliant, but a receiver could not know this.

Unfinished Business. There seems to be a paradox here:

- Reducing sidelobes should reduce sensitivity to distortion because each modulated tone extends over a narrower (and therefore less distorted) band.
- Shaping reduces the sidelobes.
- Yet shaping increases sensitivity to distortion!

The usefulness of method 3, in particular, depends on the resolution of this: Would such shaping in the transmitter make the task of an equalizer easier or harder?

6.5 DUMMY TONES TO REDUCE OUT-OF-BAND POWER?

If a "dummy" tone at the edge of a band is modulated with weighted combination of the data on a number of adjacent tones, the stopband attenuation can be increased by creating an extra spectral zero. This idea seemed to have promise for controlling ADSL leakage into the voice band and for digging spectral holes in the ham radio bands (see Sections 3.7 and 10.1.5), but it has two serious problems:

1. The method works well only if the spectral zeros are separated by $N\Delta f$. If, as in G.992 ADSL, they are separated by $N\Delta f/(N + \nu)$, the peaks of the dummy tone may correspond to the zeros of a tone it is trying to cancel, and very little cancellation can be effected.

2. Many adjacent tones may have to be weighted and combined to get any useful extra attenuation, and this may cause a very undesirable spectral peak at the edge of the band.

7

OTHER TYPES OF MCM

As we saw in Section 6.3, the big problem with DMT and OFDM—and any other MCM system that uses rectangular pulsing of sinusoidal carriers ("tones")—is the sidelobes. In summary, the high-level sidelobes of DMT:

1. Increase the sensitivity to channel distortion.
2. Increase noise enhancement in all linear equalizers and therefore make a guard period essential.
3. Make the problem of digital RFI cancellation (see Section 10.3) much harder.
4. Increase ICI and ISI at the band edges of an FDD system (this is really another manifestation of effect 1: for the wideband DMT signals, the sharp cutoff of the filters is an extreme form of channel distortion).
5. Increase the sensitivity to frequency offset [Armstrong, 1998]; this is not important in DSL systems, but it is very important in broadcast wireless systems, which use unmatched up and down conversion stages.

A guard period—and its most common form, the cyclic prefix—does not reduce the sidelobes in any way, but it does ameliorate their effects; it helps with effects 1, 2, and 4, and a shaped cyclic prefix helps a little with RFI cancellation. The 8% loss of capacity incurred in ADSL, however, represents approximately 1.5 dB loss in margin. Purists might consider such a big loss unacceptable, but a combination of sidelobe reduction with a very short guard period (<3% perhaps?) that allowed the amplitude spectrum of the SIR to approximately match that of the line—and thereby reduced the noise enhancement of the equalizer—could be very useful.

Most of the (non-DMT) systems discussed in this chapter use one or more of the following methods to reduce the sidelobes:

- Frequency-domain spreading (i.e., modulating each symbol onto a weighted sum of tones, which then constitute a "subcarrier"); this is often described as maintaining orthogonality on a distorted channel.

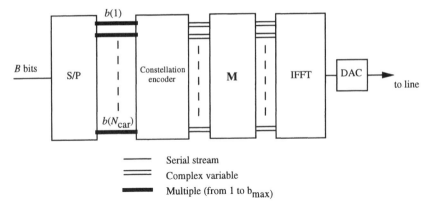

───── Serial stream

Figure 7.1 Frequency-domain spreading by matrix transformation of the IFFT input.

- Filtering of the separate parallel constellation-encoded symbols by means of a weighted sum over several intervals.
- Time-domain shaping of each symbol.

7.1 FREQUENCY-DOMAIN SPREADING

Figure 7.1 shows the essentials of this method. The memoryless matrix $\mathbf{M}$, which is simply the transparent identity matrix if no spreading is used, generates a set of basis functions ("sub-carriers") as weighted sums of sinusoids. Very general forms of the method have been described as vector coding [Kasturia and Cioffi, 1988] and structured channel coding [Lechleider, 1989], but we will confine our attention to simpler forms in which each row of $\mathbf{M}$ has at most two off-diagonal terms.

7.1.1 Frequency-Domain Partial Response

[Schmid et al., 1969] showed that partial-response coding and shaping, which is well known in the time domain (e.g., [Lender, 1964] and [Kretzmer, 1965]), can be applied with almost perfect duality in the frequency domain. The partial-response transfer functions that were of most interest in the time domain[1] were $(1 + D)$ and $(1 - D^2)$, where D is the symbol delay operator, but the frequency-domain function of most interest to us now[2] is $(1 - \Delta)$, where Δ is the frequency shift operator. This rather imprecise definition will become clearer with an example: A 7×7 matrix[3] $\mathbf{M}$ becomes

---

[1] Kretzmer called these partial response classes I and IV, and the names have been used since.
[2] In Section 11.1 we consider $(1 - \Delta/2 - \Delta^2/2)$.
[3] The exemplary size is not a power of 2, in order to emphasize that the operation is performed only on the used tones.

$$\mathbf{M} = \begin{bmatrix} 1 & -1 & 0 & 0 & 0 & 0 & 0 \\ 0 & 1 & -1 & 0 & 0 & 0 & 0 \\ 0 & 0 & 1 & -1 & 0 & 0 & 0 \\ 0 & 0 & 0 & 1 & -1 & 0 & 0 \\ 0 & 0 & 0 & 0 & 1 & -1 & 0 \\ 0 & 0 & 0 & 0 & 0 & 1 & -1 \\ 0 & 0 & 0 & 0 & 0 & 0 & 1 \end{bmatrix} \tag{7.1}$$

NOTES:

1. The only departure from duality is that time-domain partial-response encoding is continuous in time, whereas this encoding begins anew with each symbol.

2. Time-domain encoding must go forward in time, but frequency-domain encoding can go either way. Equation (7.1), for reasons that will be apparent later, shows the highest used subcarrier as a pure tone and the encoding proceeding downward thereafter. This will, somewhat arbitrarily, be designated as $(1 - \Delta)$ coding. Coding proceeding upward will be designated as $(1 - \Delta^{-1})$.

The subcarrier that is used for data symbol a_m is, in terms of the normalized variable $x = t/T$,

$$C_m = \exp(j\,2\pi mx) - \exp[j2\pi(m+1)x] \tag{7.2}$$
$$= -\exp[j\,2\pi(m+\tfrac{1}{2})x]2j\sin(\pi x) \tag{7.3}$$

which is offset by $\Delta f/2$ from the conventional tones, and shaped by a half-sinusoid. The spectrum of this subcarrier is shown in Figure 7.2 for $m = 24$, together with the spectrum of a rectangularly pulsed tone 24. It can be seen that the sidelobes of the partial-response subcarrier are much suppressed.

Sensitivity to Distortion. Figure 7.3 shows the c_1 coefficients, as defined in (6.22b), compared with those for a rectangular envelope; the sensitivity to distortion—particularly to the large terms (i small)—is greatly reduced.

Receiver Processing. The simplest method of detection would be a frequency-domain equivalent of a DFE: Make a decision on one tone, and subtract its effect from the next. Because the feedback tap for partial response is unity, however, this could lead to severe error propagation. This can be avoided by precoding[4] in the transmitter, and detecting a 2^k-level signal by slicing a $(2^{k+1} - 1)$-level eye (see, e.g., [Bingham, 1988]). The big disadvantage of frequency-domain partial response (FDPR) is the 3-dB loss in margin incurred

---

[4] Originally described in [Lender, 1964] and later generalized in [Tomlinson, 1971] and [Harashima and Miyakawa, 1972].

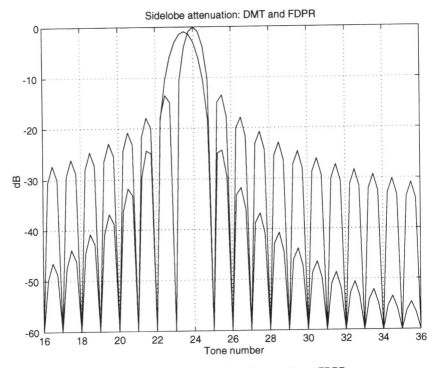

Figure 7.2   Spectra of tone 24 with and without FDPR.

by the partial-response encoding and the DFE. This loss can be retrieved by Viterbi detection[5] or by the error detection and correction algorithm (EDCA) described in [Bingham, 1988], but it not clear whether either is compatible with trellis coding. FDPR is considered again in Sections 11.1 and 11.6.

7.1.2 Polynomial Cancellation Coding

Polynomial cancellation coding (PCC) [Armstrong, 1998], in its simplest form, groups the tones in pairs and applies a $(1 - \Delta)$ operation to the pair. It achieves the same low sidelobes as FDPR but reportedly, has lower sensitivity to frequency shift and channel distortion caused by multipath. It uses only half the possible number of subcarriers, however, so its data rate is only one-half that of all the other methods. The simplest form is of interest only as an introduction to PCC with time overlap (see Section 7.3.3).

---

[5] See [Nasiri-Kenari et al., 1995] and references therein for more ideas on the use of Viterbi detection of partial-response coding.

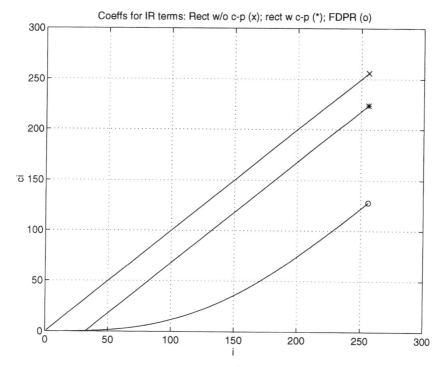

Figure 7.3 Coefficients of h_i^2 for FDPR correlation.

7.2 FILTERING

Early systems used filters that completely separated the subbands in order to maintain orthogonality, with a consequent loss of bandwidth efficiency. Several authors (see [Filt1]–[Filt5]) then described multiple staggered QAM (MSQAM), in which the filters had only 3-dB attenuation at the crossover points, thereby maintaining full bandwidth efficiency. One disadvantage is the complexity of the filters, but nevertheless, interest and progress in this method has continued (particularly for wireless broadcasting); the use of filter banks derived from those originally developed for transmultiplexers has reduced the complexity almost[6] down to that of the unfiltered approach (see [Filt 6,7]).

Another disadvantage is the extra latency caused by the filters. If a system has a maximum allowable latency,[7] the maximum symbol duration is the latency divided by the number of symbol periods spanned by the filter. Reducing the symbol duration by this factor effectively widens the filter bandwidth and

---

[6] I apologize for using such an imprecise word, but (1) I have seen conflicting estimates of the complexity, and (2) measures of complexity differ enormously depending on whether the implementation is by DSP or ASIC!

[7] For ADSL it is 1.5 ms, but for broadcast there is no limitation, which probably explains the greater interest in this method for DAB and DTV.

thereby increases the sensitivity to channel distortion. I have seen no comparisons of the relative performances of filtered and unfiltered systems when the latency is fixed.

7.3 TIME-DOMAIN SHAPING

Time-domain shaping is really just a way of FIR filtering; variations include:

- The shaping may be applied just to the cyclic prefix—a minor modification to DMT discussed in Section 6.5—or to the whole envelope pulse of duration T (Sections 7.3.1 and 7.3.2).
- The "whole pulse" shaping methods are either synchronized (Section 7.3.1), in which all the $2N_{car}$ (real plus imaginary) inputs to the modulator(s) change simultaneously, or staggered (Section 7.3.2), in which half of the inputs are delayed by $T/2$.

7.3.1 Whole Pulse Shaping with Synchronized Inputs

The advantage of the synchronized method would be that a cyclic prefix could still be used. The disadvantage would be that the only pulse shape that can achieve orthogonality *and* zero loss of margin is a rectangle: that is, no pulse shaping.[8] [Slimane, 1998] described a synchronized system that maintained orthogonality but appeared to lose 3 to 6 dB of margin. We do not discuss such methods further.

7.3.2 Whole Pulse Shaping with Staggered Inputs: SMCM

The systems described by [Mallory, 1992], Chaffee[9] and [Vahlin and Holte, 1994], were natural extensions to pulses of finite duration of the filtered systems of [Chang, 1966], [Saltzberg, 1967], and [Hirosaki, 1981]. They were also natural attempts to correct the problems of FDPR: namely, the 3-dB loss and the fact that the transmit power envelope has nulls. The Mallory and Chaffee (M & C) pulses were limited in duration to one symbol, and were raised-cosine shaped; Vahlin and Holte (V & H) showed pulses of one, one and a half, and two symbols duration,[10] and the shapes were derived from prolate spheroids that jointly minimized distortion and out-of-band energy. The sidelobes generated by M & C's raised-cosine pulse fall off very fast ($\alpha \cdot 1/n^3$ for n large), and those for V & H's pulse of duration T are very similar; we discuss only the simpler M & C pulses here.

---

[8] I have not seen this proved, but I am fairly certain that it is true.
[9] Unpublished paper and private conversations.
[10] They called them two, three, and four symbols, but their T was half of the symbol period used here.

For greatest immunity to noise the *raised-cosine* shaping should be split equally between transmitter and receiver. It is therefore more informative to define the pulse as beginning at $t = 0$ and refer to it as *half-sine-squared* shaping, which is split into half-sine × half-sine.

NOTE: If the shaping is split thus the transmit spectra for M & C pulses are the same as for FDPR.

Implementation. The figures in the earlier papers (Saltzberg, etc.) showed what might be called *alternating staggering*; this suggests a method of implementation for SMCM. The real inputs to the even-numbered channels and the imaginary inputs to the odd channels (destined to the even cosines and odd sines) form one set; they are modulated to generate a time-domain sequence x_1, which is then shaped and passed undelayed to the output. The other inputs (destined to the odd cosines and even sines) form a second set; they are modulated to generate x_2, which is shaped and delayed by $T/2$. This would, however, require two IFFTs with $8N \log N$ real multiplies each. Clearly, a first priority is to condense to one IFFT.

The implementation of a transmitter was described by Chaffee. All the data are modulated in one IFFT to generate a temporary sequence x', and then the two sets of time-domain samples x_1 and x_2 are constructed by using the facts that x_1 is symmetrical about $t = T/4$ and $3T/4$, and x_2 is antimetrical. That is,

$$
\begin{aligned}
x_1(t) &= x'(t) + x'(T/2 - t) & \text{for } 0 \leqslant t \leqslant T/2 \\
&= x'(t) + x'(T - t) & \text{for } T/2 \leqslant t \leqslant T & \quad (7.4) \\
x_2(t) &= x'(t) - x'(T/2 - t) & \text{for } 0 \leqslant t \leqslant T/2 \\
&= x'(t) - x'(T - t) & \text{for } T/2 \leqslant t \leqslant T & \quad (7.5)
\end{aligned}
$$

The steps can be described succinctly as IDFT–separate–shape–delay–add, and in the receiver they are reversed: separate–delay–shape–merge–DFT.

Orthogonality. How orthogonality is achieved can be understood from Figure 7.4, which shows the simplest cosines and sines: tones 1, 2, and 3. Within the first set (even cosines and odd sines) the $\sin^2(\pi t/T)$ shaping is symmetrical about $T/2$, and orthogonality is achieved in the conventional way with integration of the products from 0 to T. Similarly, in the second set (odd cosines and even sines) the shaping is symmetrical about T, and the integration is from $T/2$ to $3T/2$. Between the sets the $\sin(2\pi t/T)$ shaping is symmetrical about both $T/4$ and $3T/4$, and the integration is zero both from 0 to $T/2$ and from $T/2$ to T.

A More Efficient Implementation of the Raised-Cosine Special Case? As

we saw in Section 7.2, FDPR using the operator $(1 - \Delta)$ shapes the time-domain waveform with a half sinusoid. Could it not, therefore, be used instead of the two $s(t)$'s to save a further $2N$ multiplies? [Mallory, 1992] describes a

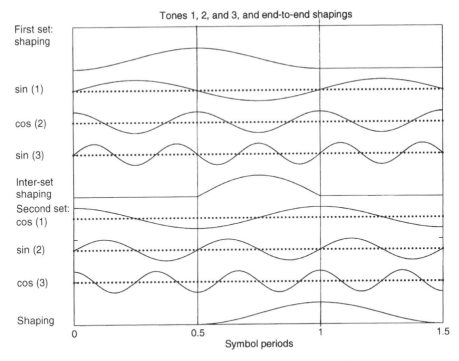

Figure 7.4   Orthogonality between tones 1, 2, and 3.

method to do this, but unfortunately I do not understand how the process in the receiver achieves orthogonality, so I can neither verify nor contradict it.

NOTE: With the first (time-domain shaping) method the modulated tones are centered halfway between the tone frequencies $n \, \Delta f$, but with the second method they would be centered around the tones. Time-domain shaping and frequency-domain spreading (partial response, correlation, etc.) are not exactly equivalent.

Performance. The primary aim of all the "other" MCM methods discussed in this chapter is to reduce the sidelobes, but one of the reasons for doing this is, of course, to reduce the sensitivity to channel distortion. SMCM certainly reduces the sidelobes, but unfortunately, it only slightly reduces the sensitivity to distortion. The "intraset" distortion (loss of orthogonality) is much reduced by the gradual turn-on of the $\sin^2(\pi t/T)$ shaping, but the "interset" distortion is only slightly reduced by the $\sin(2\pi t/T)$ shaping.

A SMCM system is therefore slightly less sensitive to channel distortion than a DMT system *without* a cyclic prefix, but much more sensitive than one with. This raises the interesting question: Is there a future for SMCM? Here are some ideas that might be explored:

- A DMT system will perform very badly if there is a major IR term beyond the reach of the cyclic prefix. This rarely occurs with xDSL, but it may be common for wireless OFDM signals subjected to multipath. The performance of SMCM on such channels should be investigated.

- In Chapter 11 we discuss frequency-domain equalization and see that its major disadvantage is the number of adjacent subcarriers that affect each subcarrier, and therefore the number of nonzero coefficients off the main diagonal in the equalizer matrix. SMCM, with its much-suppressed sidelobes, may be more amenable to frequency-domain equalization.

- *If* the equalizer problem could be solved, the much reduced sidelobes of SMCM would make it easier to (1) reduce the transmit PSD in some bands as required for VDSL (see Section 10.3), and (2) cancel RFI (see Section 10.4).

7.3.3 PCC with Time-Domain Overlap

One proposal [Armstrong, 1998] for overcoming the half-data-rate problem of the PCC system described in Section 7.1.2 is to transmit the other half of the data stream on the same set of paired tones but offset in time by $T/2$. Orthogonality between the subchannels is achieved, even without channel distortion, only by using a frequency-domain equalizer. This by itself would not be a serious disadvantage, because an equalizer for the channel will be needed anyway, but it is not clear that sensitivity to line distortion is reduced.

The interleaved signals do indeed each have a sine-squared time-domain envelope like SMCM, and the intraset (i.e., among the undelayed and the delayed sets of tone pairs) distortion is almost certainly reduced, but the interset distortion may be as high as it is for SMCM: thereby requiring a large equalizer.

7.4 DISCRETE WAVELET MULTITONE

This section is contributed by:

Aware Inc.
40 Middlesex Turnpike, Bedford, MA 01730
E-mail: telecom@aware.com

(*Author's Note*: As a DMT advocate, I do not wholly agree with the conclusions in this section, but they are more balanced than much of what we have heard from either side in the last few years, so we are making progress!)

An alternative to the classic DMT system is achieved if the DFT is replaced with a discrete wavelet transform as the modulating and demodulating vehicle [Tzannes et al., 1993], [Tzannes, 1993], [Tzannes et al., 1994], and [Sandberg and Tzannes, 1995]. This approach is usually called *discrete wavelet multiTone modulation* (DWMT), although the name *overlapped discrete multitone modulation* is also used, in tribute to a key property that is discussed later.

Let us return to Figure 5.1, the simplified block diagram of a MCM system, shown there with the IFFT implementation, (i.e., as DMT). The fundamental nature of the DWMT system is exactly the same as that of the MCM (or DMT) system, shown in this figure. The only difference between the two systems is that the modulation is implemented with an inverse fast wavelet transform rather than the IFFT, where, of course, the word "fast" designates the use of a fast algorithm for calculating the inverse discrete wavelet transform. Naturally the block diagram of a DWMT receiver would employ a fast direct (or forward) wavelet transform (FWT) in the appropriate place as well. A simplified block diagram of the entire DWMT system is shown in Figure 7.5, where this difference between the two systems is seen. At the transmitter, the outputs of the constellation encoders are used to amplitude modulate the basis elements (orthonormal collection of signals) of some wavelet transform, as indicated in this figure.

The symbolism is an important factor in understanding these systems, so a review of previously defined symbols is in order at this juncture, as well as changes in symbols and precise definitions of new ones. This is done in conjunction with the block diagram of the transmitter, shown in Figure 7.5. Let us recall that the input data to the S/P converter consists of a serial TDM data stream that is divided into *frames* of B bits each. If the frame duration is T seconds, the input data rate is $R_b = B/T$. If, for example, $T = 125\,\mu s$ and $B = 256$ bits/frame, the rate is 2.048 Mbit/s.

Recall next that with the input to the S/P converter being frames of B bits each, the output of each parallel port is a symbol (*group* of bits) with $b(n)$ bits each and

$$B = \sum_{n \leqslant M} b(n) \tag{7.6}$$

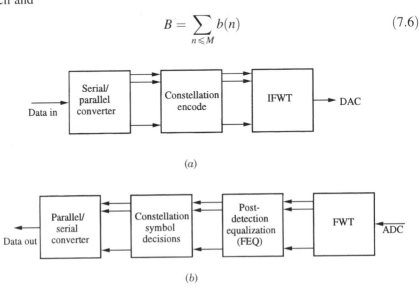

Figure 7.5 Block diagram of a DWMT system: (*a*) transmitter; (*b*) receiver.

where M (this symbol is replacing N_{car} at this juncture) is the number of groups or parallel ports, and also the number of eventual subcarriers and subchannels. The groups of $b(n)$ bits each are then constellation-encoded and modulated into the subcarriers. The number of bits in a given group is equal to the maximum number of bits that its intended subchannel can support with an acceptable symbol error rate. This is based on channel measurements made during an initialization or training period (see also Section 5.2). Some of the groups may have zero bits—channels with narrowband noise can be avoided that way.

Now let s_i^m represent the symbol in group m that came from frame i (s_1^2, for example, would denote the symbol that originated in the second group of the first frame of the serial data sequence and is headed to the second subchannel). Now we let $\nu_i^m(t)$ denote the bandpass analog signal in channel m whose origin was frame i. If this signal is sampled, it would be denoted as $\nu_i^m(l)$, where l is the sampling instance. All the $u(t)$ signals are added in time to produce the composite signal, which is sent to the channel. Let us denote by N the number of samples of this composite signal. In the DWMT case $M = N$, whereas in the DMT case $N = M + \nu$, where ν is the length of the cyclic prefix. If the rate of sampling of this composite signal is f^s, the distance between samples is $1/f^s$, and the duration of its samples for a given frame is N/f^s.

At the DWMT receiver, the FDMed signals are demodulated using the *forward (or direct)* fast wavelet transform (FWT). Each of the several sequences of the detector outputs usually undergoes equalization before decisions are made for the channel symbols. The decoded data sequences are then converted back to a single TDM stream.

Before we proceed to a detailed analysis of the DWMT system, we shall need some background on wavelets. Knowledge of wavelet transforms at the level of [Burrus et al., 1998] would be quite helpful for complete understanding of what follows. We proceed with the assumption of this knowledge. Even so, what we sketch below is also a minimal review of what is needed for the explanation of DWMT.

Wavelet theory can be developed using various approaches, the most common of which is to consider the wavelet transform (actually a series) as an *expansion*. We start our wavelet discussion here with this approach—it is the foundational approach of the theory. However, we quite rapidly move on to a second approach, the *filter bank* approach, which views the expansion as the sum of the outputs of a bank of filters, whose impulse responses are related to the basis elements of the expansion. The filter bank "realization" of a wavelet expansion is more useful in the implementation of the expansion. In presenting this filter bank approach, we concentrate our discussion on a special subcase: that of filter banks with the property of *perfect reconstruction*. The requirement that the filters of the bank possess this property is equivalent to the requirement that the wavelet basis is *complete*, which leads to the property that an expanded function is *equal to its expansion*. Within this subcase of filter banks with the perfect reconstruction property we will rapidly zero in on a special case, the *cosine-modulated filter banks* (CMFBs) [Koilpillai and Vaidyanathan, 1992]. The

CMFBs appear to be the easiest vehicles for fast implementation of the wavelet expansion, although not trivial by any stretch of the imagination.

With the two paragraphs above serving as prolegomena, we start our discussion with the expansion approach. Given a signal $y(t)$ that meets certain conditions (we omit the details of "expandability" here), it can be represented as

$$y(t) = \sum_k \sum_j a_{j,k}\, \psi_{j,k}(t) \tag{7.7}$$

where j and k are both integer indices, and the ψ's are the wavelet expansion functions that usually form an orthogonal basis. The ψ's are given by

$$\psi_{j,k}(t) = 2^{j/2}\, \psi(2^j t - k) \tag{7.8}$$

where $\psi(t)$, the function that is parameterized in two dimensions (shifts and scalings) to create the basis, is called the *generating* wavelet or *mother* wavelet. Quite obviously, it is the mother wavelet that completely specifies the wavelet system. There are many wavelet systems (orthogonal bases) generated by picking some mother wavelet that satisfies some desired condition or restriction. If the wavelet system is not only an orthogonal but also an orthonormal basis, the coefficients of the expansion (7.7) would be the inner products of $y(t)$ with the basis elements, as is always the case with orthonormal bases.

With the foregoing brief discussion of wavelets in mind, it becomes easier to see that the outputs of the constellation encoders would amplitude modulate the wavelet basis, and this would be accomplished by taking a fast discrete inverse wavelet transformation of the wavelet basis chosen for the system, with the eventual *forward* discrete wavelet transformation performed at the receiver, as shown in Figure 7.5. The expression for the signal in each subchannel would involve the information signals multiplying the corresponding basis element in the wavelet expansion, and this would presumably place the spectrum of the down-sampled signals at the appropriate channel subbands.

Well, what wavelet basis is used in the DWMT system? This can be specified by giving the mother wavelet of the expansion, since it alone completely describes the entire system. So what is the mother wavelet used in the DWMT system? How does this single mother wavelet, and the resulting orthonormal basis that it creates, accomplish spectral shifts in the system? And what do the spectra look like in the subchannels?

We are not about to give the reader the wavelet basis at this point, since the approach we will take in outlining implementation of the DWMT needs the filter bank point of view, as mentioned earlier. This point of view visualizes the wavelet basis elements as the impulse responses of filters, so instead of searching for ways to find the basis elements, one seeks to find filters with impulse responses that are the basis elements. Both approaches are considered important in understanding wavelets. The reason that we must turn to the filter bank point of view is that most of the designed DWMT systems are based on Mallat's algorithm [Mallat, 1989a, 1989b] for the structure of filters and up-samplers/down-samplers used to calculate the discrete wavelet transform, and this algorithm is based on filter bank theory. But besides providing efficient computational techniques for wavelet implementation, this theory also gives many valuable insights into the construction of wavelet bases, as well as some of the deeper aspects of wavelet theory.

At this point, then, we briefly review the theory of *M-band multirate filter banks* (a.k.a. *discrete basis functions with arbitrary overlap*) as given in [Malvar, 1992] and Chapter 8 of [Burrus et al., 1998]. These structures allow a signal to be decomposed (analyzed) into subsignals, usually at lower sampling rates, and later recomposed (synthesized) at the original rates. We limit our brief discussion to the special case of perfect reconstruction filter banks, for which the reconstituted signal is a delayed replica of the original.

Generally, an M-band multirate filter bank is specified by two collections of filters: the M analysis filters $h(k)$ and the M synthesis filters $f(k)$, $(k = 0, \ldots, M-1)$. All these filters are FIR filters with length gM, where g is the genus or overlap factor.[11] The M filters $h(k)$, which are shown in block diagram form in Figure 7.6, are chosen to approximate the ideal frequency response shown in Figure 7.7; they partition the frequency domain into M subbands (channels) of interest.

A signal $x(n)$ is passed through the bank of analysis filters $h_i(n)$, and the M output signals are *decimated* (down-sampled) by a factor of M; that is, only every Mth sample is retained. If the input signal $x(n)$ is transmitted at a rate R, the filter outputs are transmitted at the rate R/M, hence the name *multirate filter bank*. When the decimation ratio is the same as the number of filters M, the filter bank is called a *critically decimated system*.

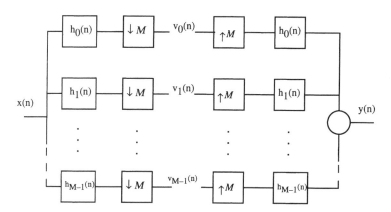

Figure 7.6 Critically decimated M-channel filter bank (analysis/synthesis).

---

[11] This terminology is established, but it is misleading; $g = 1$ means zero overlap.

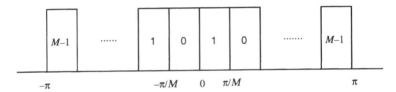

Figure 7.7 Ideal frequency response of an M-band filter bank.

To resynthesize the M signals back into one signal, each of them is first up-sampled by inserting $M-1$ zero between its samples and then passed through the synthesis filters $f_i(n)$. The M signals of this up-sampling-and-filtering operation are then added to yield the reconstructed signal $y(n)$. Under perfect reconstruction this signal is a delayed version of the original $x(n)$, that is,

$$y(n) = x(n - n^0) \qquad (7.9)$$

A sufficient condition for perfect reconstruction [Vaidyanathan, 1992] is that the analysis and synthesis filters satisfy

$$h_k[n]f_k[n + lM] = M\delta[k - k']\delta[l] \qquad (7.10)$$

It is of interest for our discussion to consider a subclass of these filter banks for which the synthesis filters are time-reversed versions of the analysis filters,[12] that is,

$$f_k(n) = h_k(L - n - 1) \qquad (7.11)$$

where $L = gM$ is the length of the filters. Under this restriction on the synthesis filters, (7.10) becomes

$$\sum_n h_k(n)h_k(L - n - 1 + lM) = M\delta(k - k')\delta(l) \qquad (7.12)$$

That is, the filters are orthogonal under shifts by M. Such a filter bank is called *para-unitary* [Vaidyanathan, 1992] by analogy with unitary matrices. This important subclass includes the wavelet transform [Steffen et al., 1993]; that is, the impulse responses of the filters make up a wavelet basis. That is why such a filter bank is often referred to as an M-band *wavelet transform*, and the analysis and synthesis filter banks as the *direct and inverse transforms*, respectively.

The DWMT system uses such a filter bank, and the definition of the analysis filters, which in turn specify the synthesis filters, is the key to implementation of

---

[12] Malvar considered the general case of the f complex, so that h is its complex-conjugate time reverse, but our f values are real.

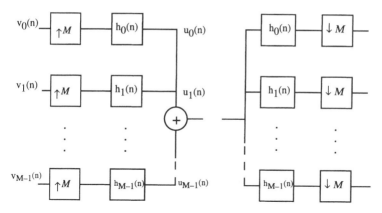

Figure 7.8 *M*-band implementation of the wavelet transform (synthesis/analysis).

the DWMT system. Of course, when the filter bank is actually used in the system, the up-sampling and the synthesis filters (which implement the inverse wavelet transform) are placed in the transmitter, with the corresponding analysis filters and the down-sampling in the receiver. This arrangement is shown conceptually in Figure 7.8. Quite clearly, the arrangement creates the needed inverse wavelet transform in the transmitter and the direct one in the receiver.

Returning now to our discussion of the filter bank, how are the analysis filters constructed using the foregoing conditions? The method for doing so is quite complicated. Filter bank design typically involves optimization of the filter coefficients to maximize some goodness measure, subject always to the perfect-reconstruction constraint of (7.12). In such a constrained nonlinear programming problem, numerical optimization leads to local minima, and the problem gets very messy when there are very many subbands, as can be the case in ADSL systems, for example.

To ease these difficulties, we usually try to impose additional structural constraints on the filters, constraints that may lead to simpler designs. The most popular such constraint is that in (7.10), all analysis (and synthesis) filters are obtained by sinusoidal modulation of a single low-pass prototype analysis (synthesis) filter. This is the key idea behind the well-known cosine-modulated filter banks, which we take up later. It is interesting to note, though, that whereas under the "expansion" approach to wavelets, one must come up with one signal—the mother wavelet. Under the filter bank approach, the problem has been reduced to finding a single filter—the prototype filter. This low-pass prototype filter is sometimes called a *window*, although it differs from the usual time windows by having the overlap property.

For such cosine-modulated filter banks (CMFBs), the synthesis filters are given by [Malvar, 1992, p. 107]

$$f_i(n) = h(n) \cos\left[\left(i + \frac{1}{2}\right)\left(n - \frac{L-1}{2}\right)\frac{\pi}{M} + \Phi_i\right] \qquad (7.13)$$

for $i = 0, 1, \ldots, M-1$, and $n = 0, 1, \ldots, L-1$, where M is the number of bands and L is the length of each filter. The filter $h(n)$ is the prototype filter we mentioned above. So we finally arrived at the point where the desired wavelet transform can be deployed by coming up with one filter, the prototype filter.

In the above equation, the parameters Φ_i are the relative phases of the modulating cosines. If we take $L = gM$, then according to Malvar, in order to have perfect reconstruction, the phases must obey

$$\phi_i = (i + \tfrac{1}{2})(g + 1)\tfrac{\pi}{2} \tag{7.14}$$

If these phases are placed into the expression for the synthesis filters above, the synthesis filters become (for $L = gM$)

$$f_i(n) = h(n)\sqrt{\frac{2}{M}}\cos\left[\left(i + \frac{1}{2}\right)\left(n + \frac{M+1}{2}\right)\frac{\pi}{M}\right] \tag{7.15}$$

where $i = 0, 1, \ldots, M-1$ and $n = 0, 1, \ldots, gM-1$. The term with the radical is there for normalization purposes (don't forget that the impulse responses of these filters form a wavelet basis).

It is shown in [Malvar, 1992] that to have perfect reconstruction with the filter bank defined in (7.15), the prototype filter $h(n)$ must obey two conditions:

$$h(n) = h(L - n - 1) \tag{7.16}$$

$$h^2(n) + h^2(n + M) = 1 \tag{7.17}$$

Using the above relations to come up with a prototype filter is not a trivial task, as has already been mentioned. We return to this issue later, after we make some additional useful comments about the overall DWMT system.

Let us assume that some prototype filter has been constructed according to some specifications and that all the various filters are in place. The sampled signal entering a typical subchannel m (which originated in frame i) has the general form

$$u_i^m(l) = p(l)\cos(\omega_m l + \phi_m) \tag{7.18}$$

where ω_m is the center frequency of the m channel and $p(l)$ is the sampled baseband pulse, common to all subchannels. Quite obviously, this $p(l)$ represents the output of the prototype filter $h(n)$. Of course, the magnitudes of these signals would be s^m, but we are ignoring that at this point, since the magnitude does not affect our discussion.

The nonzero samples of the baseband pulse $p(l)$, are in the interval,

$$0 \leqslant l \leqslant gM - 1 \tag{7.19}$$

where the parameter g is the overlap factor that was mentioned earlier. In the classical DMT system, $p(t)$ is a rectangular pulse, and its sampled version $p(l)$ has nonzero samples in the above interval with $g = 1$. This means that in the DMT case, the baseband pulses $p^m(l)$ do not overlap from frame to frame in the time domain. Thus, the values of $p_i^m(l)$ do not extend beyond the start of the next frame interval [i.e. these values do not overlap with the values of $p_{i+1}^m(l)$]. However, in the DWMT case $g > 1$, and the baseband pulse originating in frame i overlaps those originating in frames $i-g+1, \ldots, i-l$ and frames $i+l, \ldots, i+g-l$. It is this property of baseband (and passband) pulses overlapping in the time domain that resulted in the second name (overlapped discrete multitone modulation) of the DWMT system.

The $p(l)$ can also be thought off as a time window. In the DMT case this window is a rectangular pulse whose spectral density is a sinc [$\sin(x)/x$ type of shape], whereas in the DWMT case, the shape of the time window is yet to be discussed, and so is the shape of its spectral density. However, it should be emphasized that the window in the DWMT case is not like the usual time windows—it has the overlap property, a key difference from DMT. We can guess at this point that its spectral shape will probably be narrower (more confined within the subchannel bandwidth) since the baseband pulses last longer in time, and this will turn out to be a major advantage of the system. We should add at this point that wavelet transforms that have the overlap properly are in a class of transforms often referred to as *lapped transforms* [Malvar, 1992], which appear to be finding a lot of applications in signal processing.

The frequencies for the tones in (7.18) for the DWMT system are given by

$$\omega_m = \left(m - \frac{1}{2}\right)\frac{\pi}{M} \tag{7.20}$$

These frequencies are distinct and equally spaced. It should be noted that in the DMT case, the tone frequencies are given by

$$\omega_m = 2\frac{m}{2}\frac{\pi}{M} \tag{7.21}$$

These tones are in quadrature pairs (except for the two tones at the ends of the transmission band), and $(M/2 + 1)$ distinct, equally spaced frequencies are used. Quite clearly, the width of the subchannels in DWMT is half that in DMT, since the DMT channels can pass two signals each (using a sine and cosine carrier), whereas this is not possible in DWMT—we only have cosine modulated signals emitted from the synthesis filters. The narrower subchannels result in better performance in the presence of narrowband noise.

The last issue to address in the description of the DWMT system is the design of the filters. This issue is still quite complex and the subject of many current investigations. Conditions such as (7.12) do not lead to a unique solution; other constraints must be imposed, and even then, there is usually a myriad of filters that would be acceptable. There are even techniques that lead to *nearly perfect reconstruction* filters, which may be acceptable in certain applications. In some applications, the orthogonality condition may not be important and it is not imposed.

If we stick to CMFB filter banks, then we have to worry only about finding the prototype filter. But even that is not an easy task; many filters meet the conditions of (7.16) and (7.17). Also, if you come up with one, any changes in the number of subbands or the genus g lead to a different prototype. All these things are not to be viewed as disadvantages of the system. They give the designer more freedom to seek the prototype that leads to best results for his particular conditions of noise, ISI, ICI, and so on. But even with that, the existence of this myriad of prototypes leads to the inevitable conclusion that the DWMT is not one system but a large class of systems. That is why general comparisons between the DWMT and another system (DMT, say) are difficult—they apply only to the DWMT system defined by the prototype filter used by the investigator who is comparing them. A change in the prototype filter leads immediately to a change in the CMFB, and all the comparisons that were made are no longer applicable.

Space does not permit us to march the reader through the rather tedious details of the search for the prototype filter. There are some excellent sources of such techniques: [Vaidyanathan, 1992], [Malvar, 1992], [Nguyen, 1992], and most recently [Burrus et al., 1998]. A plethora of prototype filters (or windows) have already been found and some have been used in DWMT systems. As a first example, let us consider the prototype from [Kovacevic et al., 1989].

$$h(n) = \pm \sin\left[\frac{n\pi}{2(M-1)}\right] \qquad (7.22)$$

This window satisfies the perfect-reconstruction conditions of (7.16) and (7.17) but lacks *polyphase normalization*, the property that a dc input signal should be perfectly reconstructed with only the low-pass subband. This property is quite desirable in image coding because its absence leads to artifacts in the recomposed image.

To correct this problem, we consider a prototype filter that satisfies both the perfect-reconstruction and polyphase-normalization conditions: the one given in [Isabelle and Lim, 1990]. Its impulse response is given by

$$h(n) = \pm \sin\left[\left(n + \frac{1}{2}\right)\frac{\pi}{2M}\right] \qquad (7.23)$$

Figure 7.9 shows the magnitude spectra of two adjacent basis functions ($i = 2$ and 3) for the case of $M = 8$ and $g = 2$. Note that the spectra are the same for both filters (and indeed for all), since they are all cosine-modulated versions of the same prototype. Note also that the first sidelobe attenuation is approximately 24

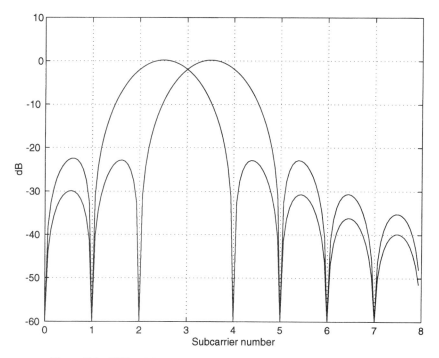

Figure 7.9 PSDs of the $i=2$ and 3 basis elements with $M=8$ and $g=2$.

dB. In general, higher spectral concentration is attained with overlap factors $g>2$. Commercially available modems typically use $g=4$ or 6.

We close this section by remarking that even though most of the discussion revolves around perfect-reconstruction filter banks, it is possible to use nearly perfect reconstruction filter banks, as well. These, when viewed as matrices, have rows that are not perfectly orthogonal to one another (as well as to themselves, at specific shifts), so they do not create an orthonormal matrix. However, if the degree of nonorthogonality is less than the noise floor of a communications system, this degree of "imperfection" in the waveform design will not affect system performance.

7.4.1 Performance Evaluations and Comparisons

We have already stressed that comparisons among DWMT and other systems (DMT, for example) are difficult, because the DWMT is not a single system but a *class of systems* depending on the prototype filter that is used in the CMFB. One could make comparisons for a specific prototype, but that would not justify conclusions on all DWMT systems. Take, for example, the comparisons between DMT and DWMT reported in [Rizos et al., 1994]. This study holds only for the specific prototype (not given in the study), which leads to the spectral characteristics of the resulting CMFB shown in the paper. The authors reported that the merits of the two systems appeared to be rather balanced. Quoting from their conclusions, "for the DFT we have reduced the SIR with the inclusion of a cyclic prefix/TDE, but this is sensitive to imperfect tap settings and it also reduces the useful data transmission time; for the CMFB, there is no particular need for pre-receiver processing, but it is sensitive to non-linear phase channels which may require optimum-combining techniques after the receiver filter banks." These comments are quite interesting, but they hold only for the prototype used (not necessarily all DWMT systems). Indeed, the prototype used in this study did not result in a set of Nyquist signals out of the CMFB, and thus the whole filter bank was not a perfect reconstruction filter bank (the authors expressed their intent to proceed to a second study with a prototype possessing these properties).

There are many prototypes in the literature, an uncountably infinite number of them. One class of prototypes presented in [Malvar, 1992, pp. 184–185], depends on a parameter that can assume any value from 0 to 1. What is presently needed is a way to classify them according to some useful characteristics so that the designer can pick a prototype that suits the specifications. In the absence of such a classification, all we can do here is a rather meager effort of discussing characteristics that hopefully hold for most prototypes, pointing out the parameters that might affect them, if we know what they are.

Latency. A key characteristic of a DWMT system is the additional degree of freedom that the genus provides, that is, the fact that the prototype filter can be longer than the frame. This property does provide us with the ability to manipulate spectral compactness, but it increases the latency in the system: a possible disadvantage in some applications. A study of latency was reported in [Wallace and Tzannes, 1995]. It is argued there that though the overlap does increase latency in the DWMT system, other factors (e.g., higher frame rate) decrease it. The latency of the DMT system is shown to be equal to two DMT frame lengths, whereas in the DWMT system it is equal to $(g + 1)$ DWMT frame lengths. However, for equal-bandwidth systems having the same subchannel spacing, the DWMT frame lengths are shown to be half the size of the DMT frame lengths, and this and some higher processing delays in the DMT system decrease the latency difference between the two. Even so, the inevitable conclusion is that if the genus is increased for added spectral containment (and the advantages that this brings), latency is also increased. The trade-off is obvious, and it is left to the designer to decide.

Equalization Considerations. Time-domain equalization (TEQ) is not needed in a DWMT system, and a cyclic prefix cannot be used. The omission of the cyclic prefix reduces the "overhead" associated with the system, since no information is transmitted by the prefix. However, the overlap property leads to the need for frequency-domain equalization (FEQ), which is performed in the receiver after the transform. The problem of interchannel interference (ICI) caused by spectral sidelobes in neighboring channels and remedies for correcting it are discussed in [Tzannes and Tzannes, 1996], [Tzannes and Sandberg, 1997], and [Heller et al., 1996]. It is asserted in these works that the equalization techniques introduced eliminate ISI and ICI. The use of a quadrature or Hilbert transform for equalization is the lower-complexity approach, since the Hilbert transform of a cosine-modulated filter bank can utilize fast algorithms. There is, of course, additional computational complexity in the use of these equalization techniques in the DWMT system (a second transform is computed at the receiver, and FIR equalizers are added on each subchannel), but it is comparable to that required for DMT for adding and removing the cyclic prefix and for performing both time and frequency-domain equalization on each subchannel output.

Complexity. The total number of operations needed per sample to implement the modulation and demodulation functions for a DWMT system is [Sandberg and Tzannes, 1995] $4(1 + g + \log_2 M)$, whereas for DMT case it is $5 \log_2 M$. It should be noted that the computational complexity required for equalization at the receiver (which is mostly predetection for DMT and postdetection for DWMT) is similar for the two systems. Now M may or may not be the same in the two systems. For a given bandwidth, if M is chosen to be the same in both systems, the two systems will have the same frame rate but not the same subchannel bandwidth because the spacing between subchannels having distinct center frequencies is twice as large for DMT as for DWMT. It is possible to select the transform size so that the subchannel spacings are the same while maintaining the same transmission bandwidth. In this case the frame rate for the DWMT is twice that of the DMT system, and the transform size M required to implement DWMT modulation/demodulation is half that for the DMT system. The increase in the frame rate of the DWMT system over that of the DMT system reduces the latency of the DWMT system, so the two systems end up with comparable latency. However, the attendant doubling of subchannel spacing can result in some degradation in performance for fixed equalization complexity. The trade-off here is obvious.

   We close the discussion of the DWMT by reminding the reader that the key new element here is use of the overlapped transform (wavelet). This provides the system with the ability to control spectral compactness and thus to ameliorate the problems caused by rectangular pulsing (particularly the high-level sidelobes). Of course, when some ills are cured, new ones may arise—it is the "there is no free lunch" principle. Still, the system adds an additional degree of freedom, which is proving quite useful in the field of data transmission.

8

IMPLEMENTATION OF DMT: ADSL

8.1 OVERALL SYSTEM

Figure 8.1 is a simple block diagram of an ADSL system. It is important to note the two stages of "splitting": from the line side the POTS and the xDSL (both bidirectional) are first separated by the low-pass and high-pass filters, and then the xDSL transmit and receive signals (unidirectional) may be separated by any combination of filters, 4W/2W hybrid, and echo canceler.

In this chapter we describe most of the components and incorporated algorithms of such a system: Section 8.2 for the transmitter, Section 8.3 for transmitter/receiver interconnection, Section 8.4 for the receiver, and Section 8.5 for the algorithms. Some of the components are generic to xDSL, some specific to DMT xDSL,[1] and some even more specific to DMT ADSL or VDSL. Some

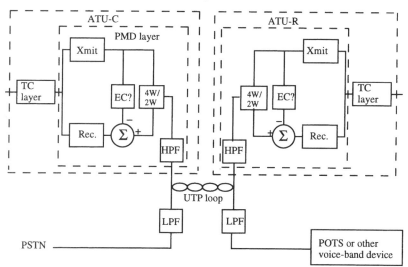

Figure 8.1 Block diagram of xDSL system.

---

[1] Strictly speaking, any modulation technique could be inserted in the transmitter and receiver boxes of Figure 8.1, but I am sure that readers will understand and empathize if I say "Perish the thought!" and do not talk about the "other" techniques.

components—POTS splitters, IFFTs and FFTs, and equalizers—however, deserve chapters or appendices of their own. In this chapter we deal with what has already been implemented; possible components of future xDSL modems—improved equalizers and RFI cancelers, and crosstalk cancelers—are discussed in Chapter 11.

System Timing. For the timing of transmission between two modems, one modem (the "master")[2] must define the frequency and phase of all clocks, and the other modem (the "slave") must lock itself to those clocks. In some early xDSL systems, the remote unit, for some mysterious and now obsolete reasons, was the master, but for all xDSL systems considered here the central unit (the ATU-C or VTU-O) is the master. The parts most concerned with timing, therefore, are the central transmitter and the remote receiver. The primary purpose of the timing "circuitry" in this pair is to reproduce the sampling clock (2.208 MHz for ADSL, 22.08 MHz for one VDSL proposal) in the receiver. This could be done by decision-aided operations on all the data-carrying subcarriers, but T1E1.4 took the easy way out and decided to reserve one subcarrier as an unmodulated pilot.

In addition to recovering the sampling clock:

1. The remote receiver must establish the symbol and superframe[3] clocks by division of the sampling clock.
2. The remote receiver must slave its upstream transmitter to this clock, and because frequency lock is thereby assured, the central receiver need establish only phase lock.
3. The PMD layer may be required to pass an 8.0-kHz network timing reference to the remote DTE. The NTR is a very precise clock that is used throughout a data network for voice sampling and CBR applications such as videoconferencing and VTOA (see Section 2.3). The NTR may also be up-sampled to get a bit clock that is locked to PRS (Stratum 1 CLK) and used for $n \times 64$ kbit/s CES.

The first two are receiver functions, which are described in Section 8.4.4. The third one requires both standardization as a transmitter function and implementation in a receiver, so it is described in Sections 8.2.1 and 8.4.3.

8.1.1 The Design and Implementation Problem

As in the design of all modems—voice-band, wireless, DSL, and so on—the basic problem is to achieve as nearly as possible the theoretical relationship

---

[2] This modem may, in turn, have to accept clocks from—that is, be slaved to—higher layers of a system.

[3] *Superframe* has a different meaning in ADSL and SDMT VDSL, but we can use the word in a generic sense here.

between data rate, error rate, and range. One way of expressing this more precisely is to define an SNR loss as the difference in decibels between the signal/unavoidable noise ratio for any loop and the achieved signal/total (i.e., unavoidable plus avoidable) noise ratio. That is,

$$\text{SUNR} = 10\log_{10}\left[\frac{\text{signal}}{\text{unavoidable noise}}\right] \qquad (8.1)$$

$$\text{STNR} = 10\log_{10}\left[\frac{\text{signal}}{\text{unavoidable noise} + \text{avoidable noise}}\right] \qquad (8.2)$$

$$\text{SNR}_{\text{loss}} = \text{SUNR} - \text{STNR} = 10\log_{10}\left[1 + \frac{\text{avoidable noise}}{\text{unavoidable noise}}\right] \qquad (8.3)$$

A measure of the state of evolution of modem design in any medium is the SNR loss achieved by an average-to-good factory-built modem. I would estimate that voice-band modems today achieve about 1 to 2 dB, but xDSL modems, which are much less mature, are probably much worse than this (4 to 5 dB?).

The main sources of unavoidable and avoidable noise in an xDSL system are as follows:

1. Unavoidable
   - Alien crosstalk
   - Kindred FEXT
   - AWGN
2. Avoidable (at least partially)
   - Kindred NEXT, whose level depends on the cable characteristics and the number of interferers, which are unavoidable, but whose effect on transmission depends on the duplexing technique used
   - RFI (AM radio and amateur radio)
   - "Impulse" noise: any noise pulses of fairly short duration that occur spasmodically and unpredictably
   - POTS signaling
   - Linear distortion (resulting, in an MCM system, in intersymbol and interchannel distortion)
   - Nonlinear distortion
   - Down/up interference: leak through FDD filters and/or echo canceler
   - Clipping
   - Quantizing noise in DAC and ADC
   - DSP round-off noise
   - Noise and/or distortion introduced by clock jitter

TABLE 8.1 ADSL Basic Numbers

	Down		
	G.992.1 and T1.413	G.992.2 Lite	Up
f_{samp} (MHz)	2.208	1.104	0.276
IFFT size	512	256	64
Cyclic prefix			
Without sync symbol		40	5
With sync symbol		32	4
Data symbol rate		4 kHz	
On-line symbol rate		4.0588 kHz	
Subcarrier spacing		4.3125 kHz	
Superframe		68 symbols plus one sync symbol	
Used subcarriers (FDD)	36–255 (note 1)	36–127 (notes 1, 2)	7–28 (note 1)
In-band transmit (PSD) (dBm/Hz)		−40	−38

Notes:
1. These numbers do not appear in any standard; they are practical recommendations to ease the filtering requirements only.
2. In G.922.2 a few subcarriers may be taken from upstream to help the downstream; the used subcarriers may be more like 33–127 and 7–25.

Equation (8.3) can then be written as

$$\text{SNR}_{\text{loss}} = \sum_{\text{all noise sources}} 10 \log_{10}\left[1 + \frac{\text{Avoidable noise}}{\text{Unavoidable noise}}\right] \qquad (8.4)$$

and each noise source should be assigned an allowable contribution (typically between 0.1 and 2.0 dB) to the total *noise budget*.

8.1.2 Numerical Details

The important numbers for ADSL are given in Table 8.1. The operative arithmetic is

$$4 \times (512 + 40) = 4.3125 \times 512 = 2208 \qquad (8.5a)$$

and

$$68 \times (512 + 40) = 69 \times (512 + 32) \qquad (8.5b)$$

Output Power Spectral Densities (PSD). As xDSL services are extended to higher and higher data rates and frequencies, crosstalk becomes more and more critical, and the transmitted PSDs have to be more tightly controlled. As discussed in Section 4.5 the specification of ADSL PSDs was tightened in Issue 2 of T1.413 to reduce crosstalk into the emerging VDSL service. The PSDs for the ATU-C and ATU-R are shown in Figures 25 and 29 of T1.413.

8.2 TRANSMITTER

Figure 8.2 is a block diagram of a DMT transmitter; it shows those parts common to Figures 2 to 5 of T1.413 Issue 2, omitting the definitions of the input data and control channels and the numbers specific to the direction of transmission. We consider each of the blocks in turn,[4] but some of them—scrambling, FEC, and interleaving—have been well covered elsewhere, and others—the inverse discrete Fourier transform and the cyclic prefix—are discussed elsewhere in this book, so I will be brief.

8.2.1 Transport of the Network Timing Reference

As shown in Figure 8.2, the NTR is an input to the mux/sync control. Information about the NTR is included in the data stream, and it is processed thereafter like any other data. Because the basic sampling clock used in an ADSL system, 2.208 MHz, is an integer multiple of 8 kHz, it might be thought that the simplest way would be to slave the whole ADSL system to the NTR. It was quickly realized during the discussions in T1E1.4, however, that a prohibitive amount of filtering would be needed in a PLL to attenuate the high-frequency components of the output of the phase detector; it would probably be nearly impossible to keep the input to the VCXO quiet enough and the resulting

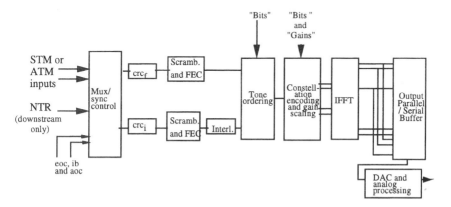

Figure 8.2 Block diagram of a DMT transmitter.

---

[4] Ideally, the length of discussion of each block should be proportional to the estimated unfamiliarity of the block to the reader, but occasionally, it will be proportional to the familiarity to the author!

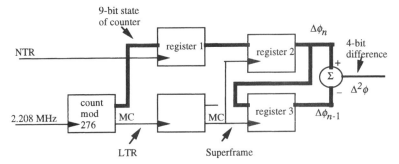

Note: MC is maxcount indication before foldover to 0

Figure 8.3 Transmission of the NTR phase information.

jitter on the derived 2.208 MHz low enough. Therefore, it was decided, surprisingly but wisely, to keep the ADSL system clocks independent of the NTR and to transmit information about any frequency offset between the local timing reference (LTR = 2.208 ÷ 276) and the NTR. The remote unit can then recreate the NTR from its own reconstructed LTR.

Figure 8.3 is a copy of Figure 9 of T1.413 Issue 2. The NTR and the LTR are both "8-kHz" clocks that are very close in frequency but have an arbitrary phase relationship. The phase of the LTR is therefore sampled on each NTR, and the value of this is sampled into register 2 at the end of each superframe. Registers 2 and 3 therefore contain the most recent and the previous phase differences (measured in units of $1/2.208\,\mu s$) between the NTR and the LTR. The second finite difference of ϕ (i.e., frequency difference) is then defined by just four bits, and transmitted as "data" in one of the downstream overhead channels. One possible circuit for recovery of the NTR is shown in Figure 8.4, and discussed, along with all other receiver timing recovery issues, in Section 8.4.3.

8.2.2 Input Multiplexer and Latency (Interleave) Path Assignment

Interleaving greatly increases the ability of the Reed–Solomon (R-S) forward error correction (FEC) coding and decoding to correct bursts of errors due to either externally generated noise impulses or internally generated clips (see Section 5.6.1), but it does increase the latency of the data. Deciding on a compromise between burst error rate and latency for each data channel is a function of the transmission convergence (TC) layer, which must combine the multiple input data channels and assign them to either the "fast" (i.e., not interleaved) or the interleaved path.

8.2.3 Scrambler

The scrambler chosen for ADSL is of the self-synchronizing type (see, e.g, [Bingham, 1988]). The descrambler for this has the effect of tripling the bit error

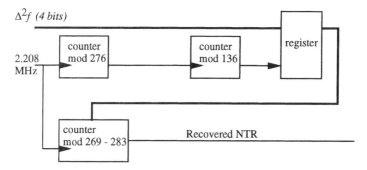

Figure 8.4   Recovery of the NTR in an ATU-R.

rate, but the general opinion seems to be that this is inconsequential if a R-S FEC is used. A DMT xDSL system must be locked to a superframe clock, and loss of that lock would affect many more receiver functions than just the descrambler, so the self-synchronizing ability of this type of scrambler brings no advantage. An additive scrambler that is reinitialized with each superframe would have been better and was proposed in [Bingham,1993], but tradition prevailed.[5]

8.2.4   Reed–Solomon Forward Error Correction

The only aspect of R-S FEC as used for xDSL that has not been covered in many books (e.g., [Berlekamp, 1980], [Clark and Cain, 1981], and [Lin and Costello, 1983]), and is the fact that the R-S codewords are locked to the DMT symbol rate. The ideal—for ease of coding at least—would be that each codeword contain exactly one symbol of data, but at low data rates this would lead to small codewords and inefficient codes, and at very high data rates it would result in inconveniently large codewords. Therefore, the number of DMT symbols per codeword (designated by S in T1.413) may be $\frac{1}{2}$, 1, 2, 4, and so on. A typical ADSL downstream transmitter at 1.6 Mbit/s would use a DMT symbol of 50 bytes and a codeword of 200 bytes; that is, $S = 4$.

One result of the integer constraint on other parts of the system is that since R-S codewords are composed of bytes, the number of bits per symbol must be an integer multiple of 8; for ADSL this means that the on-line data rate must be an integer multiple of 32 kbit/s. We look at an example of the design of an encoder/interleaver in the next section.

8.2.5   Interleaving[6]

The purpose of the combination of an interleaver in the transmitter and a de-interleaver in the receiver is to spread bursts of errors, which occur between

---

[5] I believe that there was an argument that the self-synchronizing scrambler is better for ATM transmission, but I do not remember either the argument or its source.

[6] I am indebted to Po Tong of TI/Amati for much of this section.

the two, over many codewords, and thus reduce the number of errors in any one codeword to what can be corrected by the decoder (usually half the number of redundant bytes added in the encoder). The two important parameters for an interleaver are the number of bytes[7] per codeword, N, and the interleave depth, D, which is defined as the minimum separation at the output of the interleaver of any two input bytes in the same codeword.[8] Another, system-oriented, way of defining D is as the dilution ratio of errors in a codeword. If in an MCM system with $S = 1$ (i.e., one symbol per codeword) a symbol experiences a large noise impulse such that all its bytes contain errors, then approximately $1/D$ of the bytes in each codeword at the output of the de-interleaver will be in error.

Interleavers used for xDSL are convolutional interleavers. The advantages compared to traditional block interleavers are well known: for the same N and D they require half or less than half the memory ($\leqslant 2ND$ for an end-to-end system, compared to $4ND$) and incur less than half the end-to-end delay [$N(D-1)$, compared to $2ND$]. One small disadvantage is that N and D must be coprime; that is, their highest common factor must be unity.

The interleaver/de-interleaver pair described in [Clark and Cain, 1981] and [Lin and Costello, 1983], and shown in Figure 8.5, are "triangular" convolutional. The interleave depth $D = NM + 1$ (thus guaranteeing that N and D are coprime), and the memory requirement is $(N-1)NM = (N-1)(D-1)$. The implementation is very efficient (less than a quarter of the memory needed for a block interleaver), but the constraint on D is very inconvenient for xDSL; a typical xDSL system uses $D < N$.

The interleaver defined in T1.413 was originally proposed in [Aslanis et al., 1992] and [Tong et al., 1993]. It is a generalization of the triangular interleaver in that N and D can be defined (almost) independently.[9] The interleaving rule is the

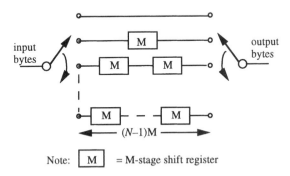

Note: $\boxed{M}$ = M-stage shift register

Figure 8.5 Triangular interleaver.

---

[7] In the literature these are often referred to as *symbols*, but we use "symbols" for a much larger grouping of bits.

[8] Readers are warned that this is the definition given in some books but not in all; careful reading is needed to reconcile the various definitions.

[9] The requirement that they be coprime is met by using only odd N values and making D a power of 2.

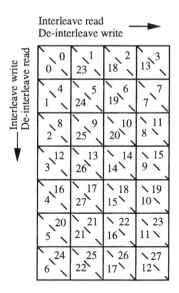

Figure 8.6 Interleave matrix for $N=7$ and $D=4$.

same as for the triangular interleaver: Each of the N bytes, B_i for $i = 0$ to $N-1$, of a codeword is delayed by $i \times (D-1)$.

Thus the first and second bytes of a codeword (indices 0 and 1) are delayed by 0 and $(D-1)$, respectively, for a net (minimum) separation of D. The interleaver can be implemented by $N \times D$ matrices that are written into by columns and read from by rows; the economy of memory compared to block interleavers is that only one matrix is used at each end: writing into one location and reading from another are performed alternately. [Aslanis et al., 1992] showed an example in which the interleave matrix is read seqentially by rows, and read—using a complicated set of rules—by columns. Figure 8.6 shows another example for $N=7$ and $D=4$ (odd and a power of 2, respectively, as required by T1.413).

NOTE: The de-interleaver is both the reverse of the interleaver (substitute "rows" for "columns," and vice versa) and the complement, in that the delay through each of them varies from byte to byte, but the total through both is constant.

A simpler way of implementing the interleaver, which requires calculating only a one-dimensional address, is with a circular buffer of circumference ND. For byte B_n

$$\text{write address} = n, \text{mod } ND \tag{8.6}$$

$$\text{read address} = \{n + ND - (D - 1) \times (n, \text{mod})N)\}, \text{mod ND} \tag{8.7}$$

It can be seen that the total memory requirement is $2ND$: twice that of a triangular pair. [Tong, 1998] describes the best of both worlds: independence of

N and D, and a $(N-1)(D-1)$ memory requirement. This is achieved—albeit at the cost of some fairly complicated programming—by reading from an address, writing to the same address, and then calculating the next address. One useful feature of Tong's implementation is that the resulting interleaving is exactly the same as that performed by the single matrix method, thus allowing for different implementations that are end-to-end compatible.

An Example of an FEC/Interleaver Combination. A typical downstream ADSL signal might have the following parameters:

Data rate	6.4 Mbit/s
Symbol rate	4.0 kBaud
Bits/symbol	1600
Bytes/symbol	200
FEC redundancy bytes	16 (8% overhead)
Correcting ability	8 bytes/symbol

If such a system experienced a large impulse that corrupted all 200 bytes in one symbol, those 200 would need to be diluted by a factor of at least $200/8 = 25$ in order to be correctable. This suggests that $D = 32$ would be an appropriate choice. The end-to-end delay for such an interleaver/de-interleaver pair would be $32 \times 200 = 6400$ bytes, which at 800 kbyte/s, would be 8 ms.

8.2.6   Tone Ordering

As we have seen in Section 5.2, subcarriers are bit-loaded in proportion to their SNR measured at the receiver, which typically, decreases rapidly with frequency. At the transmitter, however, clipping noise is impulsive in the time domain and white in the frequency domain, and would be much more damaging to those carriers that are heavily loaded if the effects of the noise were not spread out by interleaving. Tone ordering therefore arranges all the subcarriers in order of increasing numbers of loaded bits and then assigns the data in the fast (i.e., noninterleaved) path to the subcarriers in sequence followed by the data in the interleaved path to the remaining (more heavily loaded) subcarriers.

8.2.7   Trellis Code Modulation

Trellis code modulaltion (TCM) has been extensively discussed elsewhere (e.g., see [Kurzweil, 1999]), and the particular one used for ADSL is defined precisely in T1.413. There are, nevertheless, several aspects of it that are specific to MCM and need to be discussed:

- An SCM TCM system encodes and decodes the symbols of data (all with the same number of bits) in time sequence. A MCM system could do the same if it processed each subcarrier separately, but both the memory requirements and the latency of the $N/2$ Viterbi decoders would be very large. A much better idea [Decker et al., 1990] is to encode from one subcarrier to the next.

- When encoding across subcarriers the number of bits will vary from one input to the encoder to the next. This does not cause any problems, however, because only a few of the bits (just three for two successive subcarriers with the four-dimensional code used in T1.413) are involved in the coding; the other bits, whose number varies, are passed through uncoded.

- In order that each symbol can be encoded and decoded by itself without reference to the previous symbol, it is necessary to start from the first subcarrier of each symbol with the encoder in a known state and to force the encoder into that state after the last subcarrier.

- The carrier recovery circuitry or algorithms of an SCM receiver are typically only able to resolve the phase of the carrier modulo $\pi/2$. Therefore, any trellis code that is used must be $\pi/2$ rotationally invariant; the four-dimensional nonlinear code Wei code [Wei, 1984] is the best known example of such a code. A MCM system must, however, establish an absolute phase reference for every subcarrier and maintain this throughout a session. It does not therefore require the $\pi/2$ rotationally invariant property, and a simpler code could probably be used. This was proposed several times during the work on T1.413, but nevertheless, the patented Wei code was selected.

8.2.8 Pilot Tone

T1.413 specifies that one of the subcarriers (number 64) should be left unmodulated. It was argued in [Spruyt, 1997] that sensitivity to slowly varying interference would be reduced by modulating the subcarrier with random 4-QAM, but unfortunately, this good proposal came too late to be acceptable.

8.2.9 Inverse Discrete Fourier Transform

The efficient implementation of an inverse discrete Fourier transform (IDFT) as an inverse fast Fourier transform (IFFT) is discussed in Appendix C.

8.2.10 Cyclic Prefix

The form and purpose of the cyclic prefix are explained in Section 6.1. Figure 8.2 shows it graphically as the replication of inputs to the parallel-to-serial converter.

8.2.11 PAR Reduction

NOTE: The maximum permissible output signal of an ADSL transmitter is constrained by its PSD, not its total power. Therefore, an ADSL downstream signal on a long loop, which uses much less than the available 1.1 MHz of bandwidth, will transmit less power: on very long loops as much as 3 dB less. Most front-end circuitry (DAC, amplifiers, and line drivers) will be designed to handle the maximum power of approximately $+20$ dBm, so on long loops it will be operating with as much as 3 dB of "headroom", and the probability of clips will be much reduced.

The discussion in this section therefore is most relevant for short and medium loops. At least seven different methods of reducing the PAR have been suggested. Two involve setting the transmit PAR at the required low level, clipping the signal, and:

1. Accepting the errors caused; that is, allowing the application to correct them by retransmission if necessary
2. Attempting to correct the errors by an FEC with or without an interleaver

The others all involve detecting one or more potential clips in the (high-precision-wide-dynamic-range) digital summed output of the IFFT, and then:

3. Filtering the clip error(s), which are impulses, in order to move their energy up to high frequencies where the SNR is usually much lower.
4. Reducing the transmit level of the whole symbol and signaling this reduction to the receiver on an auxiliary subchannel.
5. Rotating the inputs to the IFFT by some pseudo-random multiples of $\pi/2$ and re-transforming—repeating the process up to three times if a clip still occurs.
6. Adding another signal—generated by the IFFT of calculated inputs to a small set of "dummy" subcarriers—so as to counteract the clip(s).
7. Expanding the constellation of those subcarriers that already carry data in such a way that they can be correctly detected in the receiver via a modulo operation.

The methods differ in the amount of standardization and signaling from transmitter to receiver they require:

- Filtering (3) is "transparent"; that is, it can be performed by the transmitter, and the receiver needs to know nothing about it.
- Allowing errors to occur (1 and 2), adding a predefined set of dummy subcarriers (6), and increasing the inputs to some data-carrying carriers (7) all require standardization (interface to a higher layer for 1, size of interleave buffer for 2, subcarriers to be used as dummies for 6, agreement by the receiver to use modulo detection for 7), but no signaling.
- Adding a counteracting signal on some set of subcarriers determined during training (another variation of 6) requires standardization and some minimal signaling.

- Reducing the transmit level (4) and rotating the IFFT inputs (5) require signaling with every processed symbol.

It is probable that none of these methods will be adequate by itself; a judicious combination will be needed. We discuss each of them briefly using the background information in Section 5.6.1, but the discussion will not be conclusive; many months of study and discussion among people with complementing skills will be needed before the best solution can be found. It will be obvious which are my favorites from the amount of space I devote to each method, but in order not to keep the reader in suspense (and to increase his or her reading efficiency), I will state from the start that they are numbers 3 and 7.

1. Accepting Errors and Correcting by Retransmission. It can be seen from Figure 5.1 that with a PAR of 4.0, the probability of a clip is approximately 6×10^{-5}. This means that for the 512-sample symbols used by ADSL downstream and all VDSL, the probability of a clipped symbol is approximately $1/30$. This seems like a reasonable upper demand to put on any retransmission system. Therefore, for applications that can live with a 3% retransmission rate, a PAR of 4.0 (12 dB) is achievable.

NOTE: There might be problems because the R-S codewords are synchronized with the DMT symbols; it would obviously achieve nothing if the same symbol were retransmitted.

2a. Attempting to Correct the Errors by an FEC Alone. As we saw in Section 5.6.1, a small proportion of clips will cause errors on all subcarriers loaded with more than a certain number of bits. For example, from Table 5.2 we can deduce that for a PAR of 4.0, one symbol out of 30 will have a very high error rate on all subcarriers carrying more than 7 bits. If there are more than a very few of these, it is unlikely that an FEC alone will be able to correct the errors. On the other hand, in the other 29 symbols there will be no errors and the FEC will not be needed. Clearly, an FEC alone contributes almost nothing to the correction of clip-induced errors.

2b. Attempting to Correct the Errors by an FEC and Interleaving. Interleaving has the effect of diluting each burst of clip-induced errors, and—perhaps—reducing them to a proportion that can be handled by the FEC. Analysis of the combined effects of interleaving and FEC on error rates in general, and on the required PAR in particular, is very complicated, and must be considered *unfinished business.*

NOTE: It has been argued that the FEC is intended to correct for unpredictable impulses from outside, and should not be pre-empted to deal with "predictable" events such as clips. If, however, the number of clipped symbols is kept to only a few percent, I see no reason why the FEC should not perform the double duty.

3. Filtering the Clip Impulses. [Chow et al., 1998] A clip in the time domain generates noise that is white in the frequency domain; this is potentially very harmful to those subcarriers carrying a large number of bits but totally innocuous to those carrying a small number. For minimum average deleterious effect, the clip noise should be filtered so that the signal/clip ratio (SCR) parallels the SNR.

Zero at dc. The simplest case is when the SNR decreases monotonically with frequency from dc (e.g., an SDMT VDSL system with only kindred FEXT). The clip noise should then be filtered by

$$F(z) = -0.5z + 1 - 0.5z^{-1} \qquad (8.8)$$

which

- Has a double zero at dc
- Has unity gain at $f_{samp}/4$
- Increases the total clip noise across the entire band by 1.76 dB

This is very simple to implement on the output of the parallel-to-serial converter after the cyclic prefix has been added: If the magnitude of the nth sample exceeds the predefined clip level (3.5σ is a reasonable clip level to strive for), it is clipped and half the clip noise subtracted from the $(n-1)$th and $(n+1)$th samples. If the first or last sample of the symbol must be clipped, then in order to allow each symbol to be processed completely without reference to previous or subsequent symbols, the full amount of the clip should be subtracted from the second or penultimate sample, respectively. This is equivalent to only a first-order high-pass filter, but since it is invoked only rarely, the effect on the average clip filtering is negligible.

If because of line attenuation, some of the upper subcarriers are not used, successive samples will be correlated and clips may occur in pairs. Then if both the nth and the $(n+1)$th samples exceed the clip level, they should both be clipped and half of the sum of the clip noises subtracted from the $(n-1)$th and $(n+2)$th samples. For the reasons discussed in the note at the beginning of this section, correlated clips will be rare, but modification of the algorithm to deal with them should be included.

Zero at Some Higher Frequency. If the SNR does not decrease monotonically [e.g., if NEXT from an HDSL system(s) is the dominant noise source], or if the SCR at low frequencies is of no interest (e.g., the downstream signal in an FDD ADSL system), the zero of the high-pass filter should be moved up to subcarrier n_0:

$$F(z) = -az + 1 - az^{-1} \qquad (8.9)$$

where

$$n_0 = \frac{N}{2} \cos^{-1}\left(\frac{1}{2a}\right) \qquad (8.10)$$

and for the sake of simple digital operations, a is constrained to values that can be expressed as $(2^{-1} + 2^{-m})$.

Distribution of the Clip Energy.   The filter acts on each clip separately: big or small clips become big or small filtered clips. The analysis in Section 5.6.1 of the distribution of clip energy is therefore still pertinent; the predicted levels on any subcarrier must just be multiplied by the filter transfer function.

Gain at f_{samp}.   The generalized filter of (8.9) has unity gain at $f_{samp}/4$, so all the calculations in Section 5.5.1 of the effects of the distribution of the (unfiltered) clipped energy apply unchanged to the filtered clip at subcarrier $N/4$. This allows the design process to be separated into two almost distinct steps:

1. What PAR is needed to support the required number of bits at the center of the band?
2. How should the SCCR be shaped to get the greatest margin relative to the SNR at other frequencies?

Over-Sampling.   The means for doing digital signal processing are improving much faster than those for analog processing, and oversampling and digital filtering of the transmit signal—particularly the upstream signal—are becoming more common. Oversampling is an extreme example of the situation described above—successive samples are very strongly correlated—and trying to correct for the clipping of one sample using (8.8) or (8.9) would fail because most of the adjacent samples would also be clipping. The algorithm must therefore be modified as follows.

Consider the case of $8 \times$ oversampling that is common for the ADSL upstream signal.[10] If the sample number $(8m + 1)$ clips, then $(8m + 2)$ to $(8m + 8)$ will also, and the clip must be filtered by subtracting $8a \times$ the clip from samples $8m$ and $(8m + 9)$. The resultant clip pulse $[-8a\ 1\ 1\ 1\ 1\ 1\ 1\ 1\ 1\ -8a]$ has the desired double zero at n_0, as given by (8.10). The subtraction of $8a \times$ the clip means that samples $8m$ and $(8m + 9)$ are now themselves more likely to clip, and the algorithm will perform slightly below theory; simulation is the only way to check, so an $8 \times$ oversampled ADSL upstream transmitter is included in the following simulated examples.

Some Results.   Three systems were tested by simulation. A fairly conservative PAR of 3.5 (10.9 dB) was used throughout:

---

[10] This is also discussed as *decremented oversampling* in Section 8.2.12.

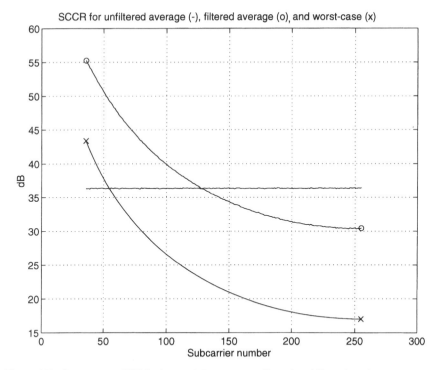

Figure 8.7 Downstream SCCRs for $a = 0.5$: average unfiltered and filtered, and worst case out of 10,000 symbols.

1. *A basic (EC ADSL down or VDSL) system:* $N = 512$, subcarriers $6-255$ are loaded,[11] and $a = 0.5$. The probability of a clipped symbol $= 0.21$. The average SCCR for the unfiltered system and for subcarrier 128 of the filtered system is 35.8 dB, but as discussed in Section 5.6.1, this is not very informative; a margin of about 18.5 dB is needed to prevent the occasional big clips and multiple clips from overwhelming the FEC. The resulting "worst-case" SCCR of 18.3 dB will support 6 bits, which is about the maximum that would be expected in the middle of the band (552 kHz for ADSL downstream, 5.52 MHz for SDMT VDSL).

   Figure 8.7 shows the average unfiltered and filtered SCCRs. As predicted by theory, the former is 35.8 dB across the band, and the latter is 35.8 dB at subcarrier 128. Below subcarrier 128 the filtered SCCR increases faster than a typical SNR (and therefore—what is most important—faster than a typical bit loading). Figure 8.7 also shows the worst filtered SCCR from 10,000 symbols (2130 clipped symbols); it is approximately 14 dB worse than the average: comparable to the 12 dB predicted in Section 5.6.

---

[11] In all cases the lower five subcarriers were omitted to allow for POTS.

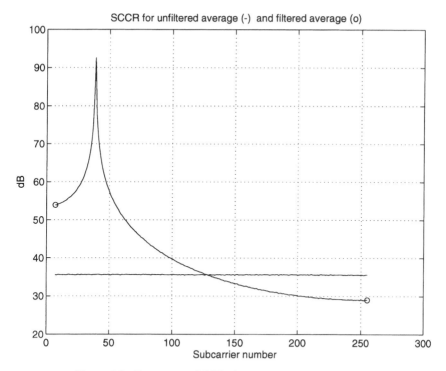

Figure 8.8 Downstream SCCRs for $a = 0.5625$ on a short loop.

2. *An FDD ADSL downstream signal on a short loop*: $N = 512$, subcarriers 36 through 255 are loaded, and $a = 0.5625$ $(2^{-1} + 2^{-4})$. The filter notch is approximately at the bottom edge of the used band $(n_0 = 38.8)$. It can be seen from Figure 8.8 that compared to the system with $a = 0.5$, at the low end of the band there is an improvement of as much as 8 dB, and at the top end there is a deterioration of about 1 dB.

3. *An $8 \times$ over-sampled upstream ADSL signal*: $N = 64$, subcarriers 6 through 28 are loaded, and $a = 0.5$ and 0.53125 $(2^{-1} + 2^{-5})$. It can be seen from Figure 8.9 that the combination of oversampling and filtering improves the SCCR across the entire band, and that tailoring the filter transfer function to match the POTS-required gap at the low end further improves the SCCR by as much as 3 dB.

An FDD ADSL downstream signal on a long loop with only subcarriers 36 through 128 loaded was also investigated, but because the total power was reduced by 3.7 dB, clips were so infrequent as to be insignificant.

4. Reducing the Transmit Level of Symbols That Would Otherwise Be Clipped. The method described in [Chow et al., 1998] greatly reduces the probability of a clip by digitally detecting a potential clip and reducing the level

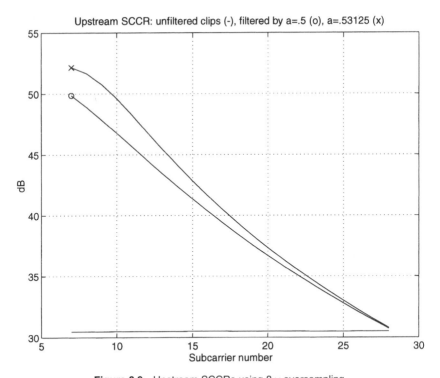

Figure 8.9   Upstream SCCRs using 8 × oversampling.

of every sample of the symbol by one of a few predefined amounts. This, of course, increases the probability of error due to external crosstalk and noise, but for a given PAR, this increase can be balanced against the decrease in clip-induced errors. The ideal balance is with a PAR of about 3.5 and the ability to deal with signals up to 6.0; this requires a range of about 4.5 dB, which can be conveniently divided into three 1.5-dB steps.

This method would require the use of two bits per symbol to signal the reduced level. Errors on these two bits would be catastrophic, but they could be made very secure by loading them by themselves on a subcarrier with a high SNR. Nevertheless, both ANSI and ITU have rejected the method as being too vulnerable.

5. Retransforming with Randomly Selected IFFT Inputs. This method was originally proposed for DMT in [Mestdagh and Spruyt, 1996], and it has been proposed for OFDM in [Müller and Huber, 1997]. In its simplest form, the method is as follows. If a sample in any symbol exceeds the clip level, the IFFT is recalculated with a new set of inputs that are related to the originals in some easily defined way but random enough that the second set of samples is independent of the first. If the new set contains a potential clip, the process is repeated.

This is not in any way a rigorous mathematical definition of the method, but the result can be simply and precisely defined. If $\mathrm{Pr}_{\mathrm{clipsymb}}$ as defined in (5.13), is the probability of a symbol containing a clip, the probability of requiring more than n IFFTs (i.e., the first n symbols all contained clips) is $\mathrm{Pr}^n_{\mathrm{clipsymb}}$ For example, for a PAR of 4.0 (12 dB) the probability of requiring more than four IFFTs is approximately 10^{-6}. The average rate at which IFFTs must be performed is

$$R_{\mathrm{IFFT}} \approx f_{\mathrm{symb}}[1 + 2\mathrm{Pr}_{\mathrm{clipsymb}} + 3\mathrm{Pr}^2_{\mathrm{clipsymb}} + 4\mathrm{Pr}^3_{\mathrm{clipsymb}}] \qquad (8.11)$$

which is typically only slightly greater than f_{symb}. If, however, the amount of RAM available for buffering and the system latency are both tightly limited, the ability to perform IFFTs at the worst-case rate of $4 f_{\mathrm{symb}}$ must be provided. Because of both the large computational penalty and the vulnerability to errors in the channel that signals the number of recalculations, the method has been rejected for ADSL by both ANSI and ITU.

6. Adding Constrained Dummy Subcarriers. There are several variations of this method (see, e.g., [Gatherer and Polley, 1997], [Tellado and Cioffi, 1998], [Shepherd et al., 1998], and [Kschischang et al., 1998]), but they all involve using only a subset of the available $(N/2 - 1)$ subcarriers for data and adding redundant signals on another subset of subcarriers so that the IFFT of the aggregate signal has a much lower PAR. The methods differ in how they choose the subsets[12]; in order of increasing sophistication they use:

1. Subcarriers at the edges of the band, where it is anticipated that the SNR will be low enough that the subcarriers would not be used for data anyway
2. Subcarriers randomly distributed throughout the band in a way that is generically optimal
3. Subcarriers distributed throughout the band so as to be optimal for the measured SNRs[13] ·

Relative to the other two, the first one has the advantage that it does not sacrifice any capacity, but it has the disadvantages that (1) a contiguous block of carriers at the edge of the band is not efficient at combating peaks, and (2) the usual reason why these subcarriers are not used is that they experience high attenuation in the channel, so the worry is that when the redundant signals are attenuted, the PAR will increase again! The other two methods have the disadvantage that they waste capacity; up to 20% of redundancy has been proposed.

---

[12] They also differ in how they calculate the redundant signals, but that need not concern us here.
[13] That is, the deduced SNR values, because the calculation is done at the transmitter where only the bit loadings are known.

7. Constellation Expansion for Some Data-Carrying Subcarriers. The following is a very brief summary of the method that is described in [Tellado and Cioffi, 1998]. Consider as a simple example an eight-level (three-bit) PAM signal that might be modulated onto one dimension of one of the subcarriers.[14] The signal points are conventionally defined as -7, -5, -3, -1, $+1$, $+3$, $+5$, and $+7$, but if instead of the point $+3$, for example, -13 were transmitted,[15] it could be correctly decoded in the receiver by a modulo-16 operation. That is, the original and "minimally expanded" sets (shown in **bold** and regular type, respectively) are

$$-15 \ -13 \ -11 \ -9 \ \mathbf{-7} \ \mathbf{-5} \ \mathbf{-3} \ \mathbf{-1} \ \mathbf{+1} \ \mathbf{+3} \ \mathbf{+5} \ \mathbf{+7} \ +9 \ +11 \ +13 \ +15$$

It can be seen that in all cases there is a change in signal value of ± 16 or, for a general L-level signal, $\pm 2L$.

To normalize the original set to unit average energy per subcarrier, they would have to be multiplied by $1/\sqrt{21}$, or, in general, by $\sqrt{3/(L^2 - 1)}$. Previously, however, we have normalized each time-domain sample to unit energy, so the normalization of the signals on each subcarrier must also include a factor $\sqrt{N_{\dim}}$, where $N_{\dim} = 2N_{sc} \leqslant N$. Then the change in baseband signal value achieved by replacing *any* original point by its minimally expanded equivalent is given by

$$\Delta_{sig} = \frac{2L\sqrt{3}}{\sqrt{N_{\dim}(L^2 - 1)}} \tag{8.12}$$

If the subcarrier onto which this baseband signal is to be modulated has a peak value (again normalized to unity) at the sample that is to be modified, the Δ_{sig} of (8.12) is also the change in the sample of the passband signal.

The basic steps of the method are therefore:

1. Identify the sample (defined as sample $n_{\max}$) of largest magnitude.
2. Identify a set of cosine or sine subcarriers—each of which defines a dimension—that have a maximum or near-maximum[16] at sample $n_{\max}$.
3. Find one subcarrier of that set that was modulated by an appropriate outer point (i.e., appropriate in that its sign was such that it contributed to the peak).

---

[14] This simplification is valid only for subcarriers with an even number of bits, for which the two dimensions can be considered separately, but the extension to the cross constellations used for odd numbers of bits is fairly straightforward.

[15] Any point defined by $(3 + 16m)$ could be transmitted, but the others would require much greater increases in energy.

[16] The reduction in PAR per step is proportional to the magnitude of the subcarrier at n_{max}, so it is advantageous to use only those subcarriers that are near a maximum.

4. Replace the point modulated onto the chosen dimension by its minimally expanded (and opposite signed) equivalent.
5. Recalculate all the samples.
6. Return to step 1 to find more points to expand so as to bring the potentially clipped signal within range.

Step 3 needs a little more explanation. Although all changes from an original point to a minimally expanded point change the signal level by the same amount, the increase in energy of the point resulting from that change does, however, vary considerably. If the original point was defined, before normalization, as $\pm(2m-1)$, for $m = 1$ to $L/2$, the increase in energy is

$$\Delta_{en} = \frac{3}{N_{dim}(L^2-1)}[(2m-1-2L)^2-(2m-1)^2] \qquad (8.13)$$

For L large the increase in energy from an inner point ($m = 1$) to its minimially expanded point is approximately $12 \times$ the average energy, but from an outer point ($m = L/2$) to its minimally expanded equivalent it is only approximately $(12/L) \times$ that energy. The power increase involved would therefore be minimized if only outer points of large constellations were expanded.

Figure 8.10 shows a 16 QAM constellation and its minimal expansion. The original points are shown **bold**; the 16 preferred points to be used for PAR reduction are shown full size; 32 of the rest of the minimally expanded set that could be used but at the expense of greater power increase are shown in smaller type; the corner 16 that should never be used because the sine and cosine subcarriers cannot have simultaneous maxima are shown in very small type.

This algorithm may require 10 or more iterations (using a different subcarrier each time) to get down to a PAR of the order of 10 dB, and Tellado reports[17] that these 10 appear to require approximately the same amount of computation as one IFFT. There is, however, little actual computation (one vector of sines or cosines per iteration) but a lot of searching and "random logic"; the algorithm will probably have to be optimized for each different DSP system.

Unfinished Business: Combining Clip Filtering (3) and Constellation Expansion (7). The main problem with the clip filtering algorithm is that the occasional large clips have enough energy that they may cause problems even when filtered. It would seem that application of just a few iterations of the constellation expansion algorithm to eliminate the peaks beyond about 5σ before filtering would be very useful.

NOTE: PAR reduction is a hot subject while I am writing this book, and everbody *may* have agreed on a method (or at least those parts of the method that have to be standardized) by the time the book appears. If so, I hope that the above

---

[17] Private conversation.

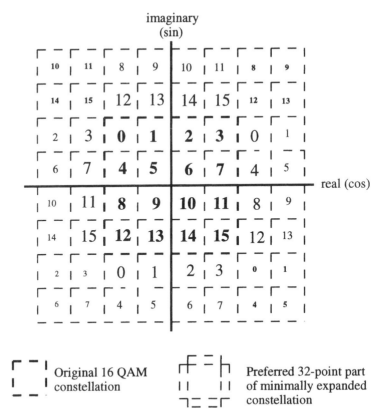

Figure 8.10 Sixteen- and minimally expanded 32-pt constellations.

discussion provides useful background to help in understanding whatever is decided.

8.2.12 Digital-to-Analog Converter

As discussed in Section 5.5, the high PAR of a multicarrier signal puts a lot of strain on many components in an MCM system, the digital-to-analog converter (DAC) among them. The conventional calculation of the number of bits required in the DAC goes as follows: For an M-bit DAC that accommodates a peak signal voltage[18] of $k\sigma$ the signal power/quantizing noise ratio, SQR, is given by

$$\text{SQR} = \frac{12 \times 2^{2(M-1)}}{k^2} \tag{8.14}$$

---

[18] This is often called the PAR of the DAC, but we must be aware of the subtle shift of meaning here; it is now the PAR of the output of the DAC, not of the MCM signal driving it; some clipping— either digitally explicit before the DAC or inherent in the DAC's input circuitry—may have already occurred.

This noise is white, so its effect will be most serious on those subcarriers that are loaded with the most bits. For a subcarrier carrying b bits, the calculation proceeds as follows:

- The SNR required for a 10^{-7} error rate is (ignoring margin and coding gain) approximately $(8 + 3b)$ dB.
- Assume that the DAC quantizing noise is allowed just 0.1 dB out of the noise budget.
- Therefore, SQR must be 16 dB greater than the required SNR.
- Each quantizing step, δ, is given by

$$\delta = \frac{k\sigma}{2^{M-1}} \qquad (8.15)$$

- Therefore,

$$\frac{12 \times 2^{2(M-1)}}{k^2} > 10^{(8+3b+16)/10} \qquad (8.16)$$

which for a reasonable maximum $b = 12$ and a typical $k = 5$ (assuming that no PAR reduction has been performed) means that

$$M > 11.5 \qquad (8.17)$$

- Rounding M to 12 bits allows k to increase to 8.0 or b to 13
- Increasing b to the maximum of 15 defined by T1.413 increases M to 13 bits

Reducing the Number of DAC Bits Required. The above calculation of the number of DAC bits is very conservative. Four methods of reducing the number have been described.

1. PAR reduction (from the $k = 5$ assumed here to about $k = 3$, thereby saving 3/4 of a bit)
2. Decremented oversampling [Flowers et al., 1998], to spread the quantizing noise over a wider band
3. Run-sum filtering to reduce the quantizing noise at low frequencies where the SNR is highest.
4. Predistortion to reduce the signal at high frequencies where the SNR is lowest

NOTE: All of these methods would be as applicable to SCM as to MCM, but they are particularly useful for MCM because of its more stringent conversion requirements.

PAR reduction has been discussed in Section 8.2.11; the others are discussed below.

Decremented Oversampling (DOS). For some DAC technologies it may be advantageous to increase the sampling frequency above that required by the transmitter (i.e., to oversample) in order to reduce the number of bits required. Oversampling is particularly appropriate for the upstream ADSL channel. The minimum sampling rate is 276 kHz, but digital oversampling by a factor of 8 (up to 2.208 MHz, the downstream sampling rate) may be convenient. If the same amount of quantizing noise were then spread over eight times the bandwidth, its level would be reduced by 9 dB, thereby saving $1\frac{1}{2}$ bits.

The simplest method of oversampling is to replicate the samples seven times. This would result in a zero of the output spectrum at 276 kHz with a "$\sin x/x$" roll-off of 3.92 dB at the 138-kHz band edge. Such replication would, however, also replicate the quantizing noise and leave most of it at low frequencies. In order to spread the quantizing noise over the whole wider band, it must be decorrelated from one oversample to the next. One (and maybe the simplest) way to do this is to decrement each successive repeated sample by a factor $(1-\Delta)$. For $M=12$ and $k=5$ as in the previous example, each quantizing step $= 5\sigma \times 2^{-11}$, so $\Delta = 2^{-8}$ allows for a very simple decrementing operation (shift right by eight and subtract) and ensures that each successive sample will move into another quantizing bracket.

Run-Sum Filtering (RSF). For the great majority of loops, signals, and interferers, the SNR decreases almost monotonically with frequency; as we have seen, the low frequencies often support 12 or more bits, and the channel is truncated when it can no longer support one. Clearly, quantizing noise is more harmful at low frequencies than at high frequencies, and ideally it should be high-pass filtered.

A very crude filter can be implemented by maintaining a running sum of the quantizing errors and shifting the quantized output by one step whenever the running sum exceeds half a step; that is:

- $[x] =$ the quantization of x to the nearest integer number of steps, $n \times \delta$
- $e = x - [x]$
- $\mathrm{rsum} = \mathrm{rsum} + e$
- If $\mathrm{rsum} > \delta/2$, then $[x] = (n+1)\delta$ and $\mathrm{rsum} = \mathrm{rsum} - \delta$
- If $\mathrm{rsum} < -\delta/2$, then $[x] = (n-1)\delta$ and $\mathrm{rsum} = \mathrm{rsum} + \delta$

This algorithm was simulated for the 12-bit DAC with a 5σ range that was recommended by (8.17). The conventional quantizing noise was, as would be expected, constant at -63 dB relative to the signal; the filtered quantizing noise was less than this over approximately the lower one-third of the band and 6 dB greater at the top of the band (subcarrier 255).

The algorithm zeros the dc component of the quantizing error, and it appears that the operation is equivalent to a first-order (single real pole) high-pass filter.

The transfer function of this filter was found from the best match to the simulated results to be

$$H_{rs} = \frac{2\sqrt{2}jf}{jf + f_N} \tag{8.18}$$

where f_N is the Nyquist frequency ($=f_{samp}/2$). The model can be justified theoretically—with the benefit of a lot of hindsight—by the following, not very rigorous, argument.

- The cutoff frequency is determined by the average rate at which the running sum approaches the reset thresholds of $\pm \delta/2$.
- Since the quantizing noise is uniformly distributed over the interval $-\delta/2$ to $+\delta/2$, it would appear that the average number of samples to reach a threshold is two: resulting in a "cutoff" at half the sampling frequency.
- The occasional adjustment of the quantized value has the effect of doubling the total quantizing power; the infinite-frequency gain of $2\sqrt{2}$ in (8.18) is needed to ensure this.

The use of this RSF algorithm for a TDD VDSL system[19] that experiences only FEXT is shown in Figure 8.11, which plots the SNR and the SQRs for a *11-bit* conventional DAC and a *9-bit* DAC using RSF. As a check on the model, it also superimposes a few points for a 9-bit DAC filtered by (8.18). It can be seen that the model fits the simulated performance very well, and—more important—the filtered SQR approximately parallels the SNR.

The effects of each of the DACs on the performance of the system (i.e., their share of the noise budgets) can be quantified as the logarithmic ratios of external noise plus quantizing noise to external noise alone:

$$dB_{loss} = 10 \times \log_{10}\left(1 + \frac{SNR}{SQR}\right) \tag{8.19}$$

These are plotted in Figure 8.12. As would be expected, the conventional DAC is better at high frequencies, but both are unimportant there anyway; the run-sum-filtered DAC (*with two fewer bits*) is much better at low frequencies, where both are important.

Predistortion (Unfinished Business). The calculation of M in (8.17) is based on $b = 12$, which is the maximum loading that will be applied. A SQR of $(8 + 3 \times 12 + 16) = 60\,dB$ is needed only at low frequencies, so the power of the full signal, which controls the amount of quantizing noise, can be reduced by digitally de-emphasizing the high frequencies. This de-emphasis, which is

---

[19] This chapter is about ADSL, but it seems appropriate to cite some VDSL results here.

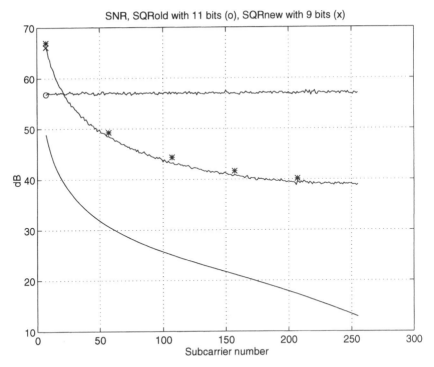

Figure 8.11   SNR and SQR for VDSL system with FEXT.

the complement of the run-sum (high-pass) filtering of the noise, could be done with a simple third-order digital FIR filter, and the output spectrum could be flattened with a three-pole analog filter. It might be possible to use the transformer and dc-blocking capacitor to perform this flattening. An even more interesting idea, which would be appropriate for the upstream transmitter, is to combine this with the pre-equalizer discussed in Section 8.3.4; that is, use the g_I values to shape the signal, and then both flatten the output PSD and pre-equalize the upstream band with the high-pass filter.

Combination of the Algorithms. If the DOS and RSF algorithms are combined (DOS first) the quantization noise is reduced *and* much of it is swept up into the unused higher part of the band. For an $8 \times$ oversampled ADSL upstream signal, a 9-bit DAC with the DOS/RSF algorithm is better than a 12-bit conventional DAC over the lower part of the band and worse over the upper part. The relative dB_{loss} values [as defined by (8.19)] incurred will depend on the shape of the SNR curve, which in turn depends on the type of crosstalk (HDSL NEXT? EC or FDD ADSL?), but a saving of 3 DAC bits—approximately $1\frac{1}{2}$ bits each from the two algorithms—relative to a conventional system can often be achieved. Run-sum filtering and pre-emphasis are probably alternatives that should not be combined; they both strive to shape the SQNR to match the expected SNR of the loop.

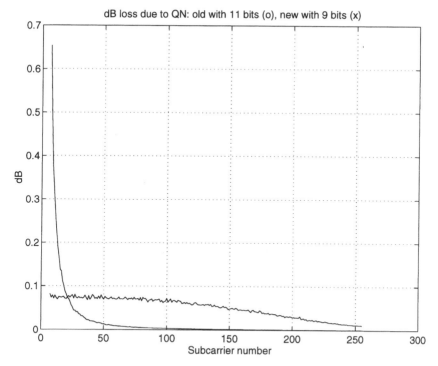

Figure 8.12 Decibel loss of conventional and run-sum filtered DAC.

8.2.13 Line Drivers

The main problem for xDSL line drivers is the required voltage swing. Even if the PAR can be reduced to 10 dB, for an average downstream output power of 20 dBm (0.1 W) into a 100-Ω line, as defined in T1.413, the peak line voltage, V_{max}, is given by

$$\frac{V_{max}^2}{R} = 1 \text{ W} \tag{8.20}$$

whence

$$V_{max} = 10.0 \text{ V} \tag{8.21}$$

Furthermore, since the line must be driven from a 100-Ω source, the driver voltage must be twice this[20]; 20 V is very difficult to get out of an integrated circuit! Similar calculations for upstream (-38 dBm/Hz across approximately 100 kHz $= 12$ dBm $= 16$ mW) would call for a driver voltage of 8.0 V. One

---

[20] Methods have been described for incorporating the line input impedance in the feedback of an amplifier and thus saving the voltage and power that are "wasted" in the driving resistance, but it is very difficult to incorporate these circuits into the 4W/2W hybrids and/or echo canceler circuits.

solution to this problem is to use two push/pull amplifiers with "inherent impedances" (ratio of voltage drive to current drive capabilities) of about 30 Ω to drive the (balanced) primary of a 1:3 transformer; this is discussed in more detail in the next section.

8.3 FOUR-WIRE/TWO-WIRE CONVERSION AND TRANSMIT/RECEIVE SEPARATION

8.3.1 Line-Coupling Transformer

The traditional line-coupling transformer is a three-port device that performs three functions:

1. Common-mode isolation (lightning protection).
2. Unbalanced-to-balanced conversion; all internal operations are performed unbalanced (i.e., referenced to ground).
3. Four wire-to-two wire conversion with (partial) separation of transmit and receive signals; if the hybrid impedance equals the input impedance of the line (seen through the transformer), there will be infinite loss from transmitter to receiver.

For xDSL these functions must be modified:

- As discussed in the preceding section, there are advantages to using balanced line drivers, so function 2 is not needed.
- The inherent impedance of amplifiers is much less than 100 Ω, so we now need an impedance transformation.
- The shunt inductance of the transformer can perform part of the high-pass filter function needed for the POTS splitter.

One possible configuration is shown in Figure 8.13.

8.3.2 4W/2W Hybrid

The first means of separation of transmit and receive signals is the 4W/2W hybrid. The maximum amount of separation that can be achieved is the return loss (RL) between the input impedance of the line and the reference impedance (perhaps hypothetical[21]) of the hybrid, and this RL must be taken account of when designing the FDD separation filters (see Section 8.3.4). For loops without bridge taps the input impedance is approximately equal to the characteristic impedance, whose variation with frequency can be well modeled by an RRC

---

[21] The 4W/2W network may not contain a reference impedance per se; it may include a transfer function, which attempts to model the reflection coefficient, and a subtraction circuit. The theoretical performance limit would be the same.

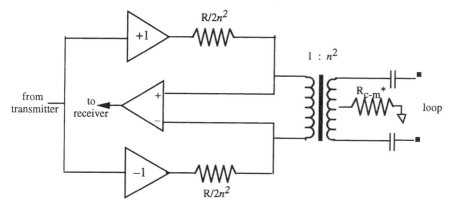

* This resistor terminates the common mode only; opinions differ on what its value should be.

Figure 8.13   Balanced line drivers and coupling transformer.

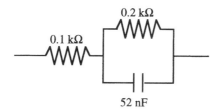

Figure 8.14   Compromise RRC model of loop input impedance.

impedance. Figure 8.14 shows an impedance that is a compromise between 24 and 26 AWG loops, and Figure 8.15 shows its RL against the input impedance of two "basic" loops defined in T1.413 and G.995: CSA 6 (9 kft of 26 AWG) and CSA 8 (12 kft of 24 AWG). These are best cases, which maintain an average RL of about 28 dB across the band.

Bridge taps near the end of a loop greatly reduce the RL at that end[22] around the "notch" frequency, which is a function of the length of the bridge tap (see Section 3.5.2). Figure 8.15 also shows the return loss of CSA 7 (another one of the test loops), which has a bridge tap right at the end; the minimum return loss is about 4 dB! A conservative design approach—at the RT at least—is therefore to assume that a bridge tap can be of almost any length and at any distance from the RT. Analysis of many different loops using the program in Appendix A suggests that a worst-case RLRT of 5 dB should be assumed across the entire band.

The situation at the CO is more controversial. Bridge taps in the feeder cable are certainly less common, but according to [AT&T, 1982] they do occur, and the test suite defined in T1.413 includes one such loop. Therefore, the conservative approach is to assume a worst-case RLCO of 5 dB also.

[22] They are almost invisible from the other end.

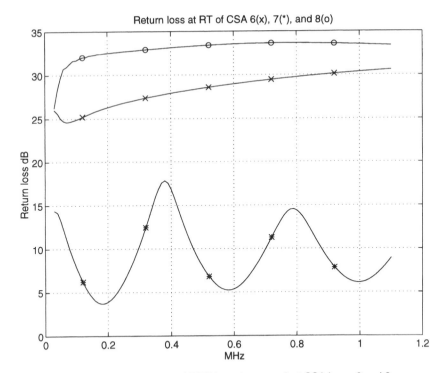

Figure 8.15   Return loss of RRC impedance against CSA loops 6 and 8.

An xDSL unit, of course, sees the loop through the second-order high-pass filter, and the RL value will be very low over a significant frequency range (far beyond the 30 kHz cutoff frequency). The RL could be improved either by incorporating an average L and C into the hybrid reference impedance, or by passing the transmit signal through an equivalent third-order active-RC transfer function and subtracting it from the reflected signal. Because of the problems with bridge taps, however, it is better—for FDD systems at least—to design the filters for worst-case RL values and live with reflections from the high-pass filter.

NOTE: It has been suggested that ADSL modems designed specifically for countries that do not have bridge taps in their cables might rely on the higher RLs and have simpler filters. To take advantage of the greater certainty about the input impedance of the loop, however, these modems would have to compensate for reflections through the high-pass filter: probably just as difficult as providing the extra filtering.

Adaptive Hybrid. The trans hybrid loss (THL) can be improved adaptively either by adjusting the balance impedance in the hybrid or by passing the transmit signal through a separate echo-emulating path and subtracting the result from the reflected signal. The latter approach seems to be preferred, but application of the method at xDSL frequencies has previously been hampered by the difficulty of implementing highly linear, controllable analog components (resistors, capacitors, multipliers, transconductance amplifiers, etc.). Recent work, reported in [Pécourt et al., 1999], appears to have solved this problem, however, and THLs >25 dB have been achieved even with the most demanding bridge taps. The next step must be the development of on-line algorithms for calculating the parameters of the emulating path.

Tuned Adaptive Hybrid. The attenuation from transmitter to receiver is provided by filter plus hybrid plus filter, where each "filter" may be the combination of analog, digital, and (I)FFT. The combined filters typically have the least attenuation around the crossover frequency, so that if the error measure used for adaptation of the hybrid is equally weighted at all frequencies, the total attenuation will have a minimum in that region. A better strategy is to weight the adaptation error more heavily in the crossover region so that the hybrid is partially "tuned." This would allow system management to partly control the crossover frequency—perhaps between 100 and 150 kHz—in response to a combination of loop lengths, traffic needs, and binder-group management (see Section 4.6.5).

NOTE: The crossover frequency must be the same for all the pairs in the binder group to avoid kindred NEXT.

8.3.3. Echo Canceler?

In Section 4.2.1 we distinguished between EC as a duplexing strategy—that is, using band 1 for both downstream and upstream and EC as an implementation tactic. We now need to consider both for ADSL.

EC as a Duplexing Strategy. T1.413 and G.992.1 state that EC is optional for ADSL, but my conclusion in Section 4.4 was that for most systems it is obsolescent. Now I will go so far as to say that allowing it in G.992 was a mistake. This does not, however, mean that ATUs with echo cancelers are obsolete. The ATU-Cs can simply turn off band 1, and both ATU-C and ATU-R can use the cancelers to assist in band separation. Because EC for ADSL is very complicated in both design and implementation,[23] but has nevertheless been well covered elsewhere (see the specialized bibliography at the end of the references) I will not discuss it here.

EC as an Implementational Tactic. Most designers of DMT transceivers have long realized that DMT does not like being bandlimited, and FDD filters *may* be

---

[23] Usually comprising three cancelers: a precanceler to protect the ADC, a time-domain canceler to remove the noncyclic part of the echo, and a frequency-domain canceler.

a big (perhaps the biggest) contributor to distortion.[24] Therefore, it was thought that the sidelobes should be removed by a simplified EC instead of filters. Methods of implementing these are proprietary, but it is clear that at the very least they must use a precanceler to protect the ADC. Prudently designed filters would probably do almost as well and would be *much* simpler.

8.3.4 FDD Filters

T1.413 specifies the out-of-band PSDs for both transmitters, but it does not specify the separating filters needed for FDD operation; they were left as vendor discretionary because their primary purpose is to protect the "near-end" receiver. The lower end of the downstream band as specified in Figure 25 of T1.413 is assumed to be that appropriate for an EC system, and extends down to 26 kHz; the specification is much looser than that needed for an FDD system. Only for the upper end of the upstream band do the two requirements meet, and then the roll-off of the ATU-R PSD, as defined in Figure 29 of T1.413 to limit XT into other xDSL systems, is comparable to that needed for FDD.

In an ATU-C transmitter some separation is achieved merely by turning off subcarriers 1 to 35; the lower sidelobes of the used subcarriers (36–255) are attenuated by the IFFT. In an ATU-R receiver the reflection of the subcarriers transmitted in the low band is attenuated by the lower sidelobes of the FFT. Similar effects can be achieved in the upstream direction by performing a double-size (i.e., 128-pt) IFFT and FFT. This requires extra computation, but it provides distortion-free filtering. It is important to note that if this is not done in the ATU-R transmitter, the filtering needed to meet the upstream PSD mask (regardless of any calculations about interference with the receive signal) is very sharp.

Figure 8.16 shows the PSDs of received signals and unavoidable noise (10 HDSL and 10 ADSL crosstalkers) for CSA loop 8, assuming that as suggested in Table 8.1, subcarriers 29 through 35 are sacrificed to a guard band. Transmitter leakage into the receiver is a major impairment in ADSL systems, but filtering (particularly analog) consumes power and distorts the signal, so it is advisable to allow leakage to be as big as possible and use as much as 1.0 dB of the noise budget. This means that leakage can be about 6 dB below the unavoidable noise.

The design of filters for FDD DMT is more complicated than for SCM because the reflected transmit signal must be considered in each separate subchannel rather than as an aggregate across the band. If the design is done carefully and prudently, however, the resulting filters should be less complex than those required for SCM because of the filtering inherent in the IFFT and FFT [smoothed as shown in Figure 6.2 and modeled by (6.18)].

It is best to consider the two filters at an ATU together and to design them by iterative modeling (using poles and zeros immediately rather than templates such as Butterworth or elliptic[25]) and analysis. The power spectrum of the signal

---

[24] The only explicit statement of this that I have seen, however, is in [Saltzberg, 1998].
[25] The poles and zeros of one of these filters could be used as a starting point for the iteration.

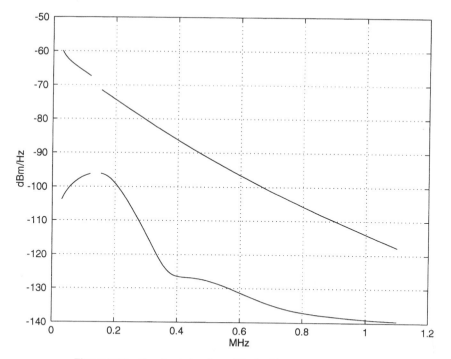

Figure 8.16 Received signals and "noise" (crosstalk) for CSA 8.

delivered to the FFT can easily be calculated from the tandem connection of the IFFT, transmit filter, and receive filter:

$$S_{ITR}(f) = S_I(f)S_T(f)S_R(f) \qquad (8.22)$$

or more conveniently, as a function of the tone number, n:

$$S_{ITR}(n) = S_I(n)S_T(n\Delta f)S_R(n\Delta f) \qquad (8.23)$$

where, for convenience, S is written for $|F|^2$. Then the interfering transmit signal appearing at the output of the FFT in subchannel m is given by the convolution

$$S_{ITRF}(m) = \sum S_{ITR}(n)S_F(n - m) \qquad (8.24)$$

and the independent parameters of each filter—preferably as few as possible [26]—can be found by iteration.

---

[26] A filter with a maximally-flat passband is ideal because it can be fully defined by its transmission zeros.

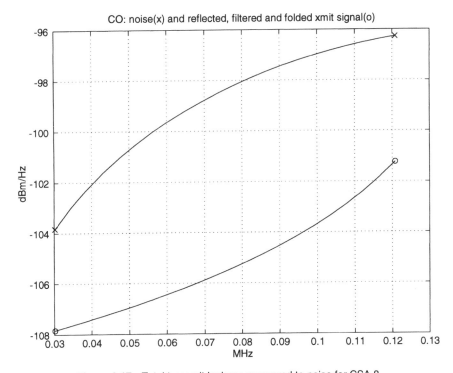

Figure 8.17   Total transmit leakage compared to noise for CSA 8.

The S_{ITRF} value of a pair of CO filters—inverse Chebyshev with 35 dB of stopband rejection—designed this way is shown in Figure 8.17 together with the unavoidable noise for CSA 8; the goal of a 6-dB "margin" is achieved. The RT filters are a little more problematic; the receive filter would be considerably simplified if the cyclic prefix were shaped in the receiver according to method 2 of Section 6.4. Design of the filters should be postponed until it has been decided whether the extra distortion associated with the shaping can be tolerated.

8.4   RECEIVER

NOTE: Some parts of the discussion in this section may seem rather vague, but at this early stage of the technology many of the details of receivers are proprietary,[27] and I can say no more.

Much of the receiver—the analog-to-digital converter (ADC), the FFT, the Viterbi trellis decoder, the de-interleaver, the Reed–Solomon decoder, and the

---

[27] Receivers are not defined in any standard; they have only to demodulate and decode a defined transmit signal.

descrambler—is the mirror image of a transmitter, and furthermore, much of it is not unique to either DMT or xDSL; nevertheless, for the sake of consistency and continuity, these components are each given a (sometimes very short) section of their own.

8.4.1 Analog Equalizer?

In many modems where adaptive equalization is needed, a pre-equalizer (fixed or switchable in very coarse steps) is used to reduce (1) the variation of attenuation with frequency, (2) the spread of eigenvalues of the signal input to the adaptive equalizer, and (3) the convergence time of that equalizer; this pre-equalizer is frequently analog. We have to consider whether such an equalizer should be used for xDSL. The conditions and conclusions are different in the ATU-C and ATU-R, so we will consider them separately.

ATU-R. The PSD of a downstream ADSL receive signal typically decreases rapidly and—ignoring dips due to bridge taps—almost monotonically with frequency. The average xDSL noise, on the other hand, is *approximately* white (NEXT increases with frequency, FEXT decreases with freqency), and the quantizing noise out of most analog-to-digital converters (ADCs) is also white. The optimum conditions for an ADC—with the SQR greater than the SNR by a constant amount at all frequencies—are thus achieved. If, however, the signal plus noise are analog equalized, the SQR at low frequencies, where the SNR and bit loadings are highest, will be much reduced. In extreme cases the ADC would need an extra two bits to achieve the same performance. The conclusion is that there should be no analog amplitude equalization[28] in the ATU-R.

ATU-C. The variation in received level across the 30 to 110 kHz received band is much less than at the ATU-R, so the considerations about SQR in the ADC are less important. On the other hand, the variation per Hertz and the resultant distortion are greater, so it might be useful to ease the task of the TEQ (Section 8.4.4) by some pre-equalization. This could be done by raising the cutoff frequency of the high-pass filter to about 100 kHz. This would have two other small benefits:

- The inductance of the transformer and the resulting distortion due to the dc current would be reduced.
- The duration of the impulse response of the total channel would be reduced.

NOTE: The slope in the high-pass response would affect the upstream transmit signal also and would have to be compensated for digitally by the g_i values (see Section 5.3).

---

[28] There might be some value in an all-pass delay equalizer; I know of no discussion of this.

8.4.2 Analog-to-Digital Converter

Some PAR reduction techniques—particularly methods 6 and 7 of Section 8.2.11—may be slightly reversed by the filtering performed by the line and/or the 4W/2W hybrid, but the PAR of a receive signal is probably not much different from that of a transmit signal. There are two conflicting factors operating here; compared with the DAC in the transmitter:

- The peak voltage-handling capability of the receiver front-end circuitry is not as important as that of the transmitter because the voltages and consumed power are much lower.
- On the other hand, bits are more expensive in an ADC than in a DAC.

A compromise decision would be to set the PAR of the ADC 1 or 2 dB higher than that of the DAC.

The high-frequency roll-off of the channel, which greatly reduces the total power of the receive signal, performs the same function for the ADC as predistortion does for the DAC. Consequently, the requirement for 11.5 conversion bits calculated in (8.17) can be much reduced; 10 bits would be ample; 9 would probably suffice; 8 would be pushing it!

8.4.3 Timing Recovery and Loop Timing

It is important to note that in a DMT xDSL transmitter the clock and subcarrier frequencies are related by integers, and because there can be no frequency shift in the channel, they are similarly related in the receiver. Therefore, recovery of the "sampling clock" in the receiver is equivalent to recovery of the "carriers." Furthermore, the only offset that the recovery circuitry has to deal with is that caused by the mismatch of two crystal oscillators: typically, less than ± 100 ppm.

There is certainly sufficient information contained within any randomly modulated MCM signal to allow recovery of the sampling clock, but early ADSL systems took the easy way out: they accepted a very small loss in data rate and dedicated one unmodulated subcarrier (n_p) for use as a pilot. If this tone is considered to be real, a feedback loop can be constructed to drive its imaginary part to zero.

The imaginary part of the complex output of bin n_p from the FFT is input to a loop filter, which, via a simple DAC, delivers a control voltage to a voltage-controlled crystal oscillator (VCXO). Ideally, the loop filter would calculate the frequency of the sampling clock to be used for conversion and demodulation of the next symbol, but because of the time required to perform the FFT, the new frequency is not available until the following symbol, and an extra factor of z^{-1} must be inserted in the loop. The design of such loops is well covered in the literature (e.g., see [Lindsey, 1972] or [Gardner, 1979]).

Specification of the Recovered Sampling Clock. The control voltage for the VCXO must be maintained for one symbol period. The permissible error in this voltage (or, rather, in the induced VCXO frequency) can be calculated as follows.

The received signal is the sum of subcarriers (n_1 to n_2) randomly modulated by $a(n)$:

$$S(k\tau) = \sum_{n=n_1}^{n=n_2} a(n)e^{j2\pi nk/N} \tag{8.25}$$

If this is demodulated (FFTed) using an offset sampling frequency of $f_{\text{samp}}(1+\Delta)/2$, the appropriately scaled output for subcarrier m is

$$Y(m) = \frac{1}{N} \sum_{n=n_1}^{n=n_2} \sum_{k=0}^{N-1} a(n)e^{j2\pi k[n-m(1+\Delta)]/N} \tag{8.26a}$$

$$\approx \frac{1}{N} \sum_{n=n_1}^{n=n_2} \int_0^N a(n)e^{j2\pi x[n-m(1+\Delta)]/N}\,dx \tag{8.26b}$$

$$= \sum_{n=n_1}^{n=n_2} \frac{1}{j2\pi[n-m(1+\Delta)]} a(n)[e^{j2\pi[n-m(1+\Delta)]x/N}]_{x=0}^N \tag{8.26c}$$

$$= a(m)(1 - j\pi m\Delta) + \sum_{n\neq m} a(n)\frac{-m\Delta}{n-m} \tag{8.26d}$$

The first term in (8.26d) is what would occur with an SCM system: the desired output and a quadrature distortion term proportional to the shift of its carrier. The second term—in-phase interchannel distortion from all the other subcarriers—occurs only with MCM.

If each of the $a(n)$ is considered to have unit power (i.e., $E\{|a|^2\} = 1$), the total distortion is

$$E^2 = (m\Delta)^2 \left[\pi^2 + \sum_{k=1}^{k=n_1-m} \left(\frac{1}{k}\right)^2 + \sum_{k=1}^{k=n_2-m} \left(\frac{1}{k}\right)^2 \right] \tag{8.27}$$

$$\approx (m\Delta)^2 \left(\pi^2 + \frac{2\pi^2}{6} \right) = \frac{4(m\pi\Delta)^2}{3} \quad \text{for } n_1 + 1 \dot{<} m < n_2 - 1 \tag{8.28}$$

Therefore, the signal/distortion ratio on subcarrier m is

$$\text{SDR}(m) = -20\log(m\pi\Delta) - 1.2\,\text{dB} \tag{8.29}$$

The tolerable value of Δ depends on the number of bits loaded onto each subcarrier, but for a typical FDD ADSL signal, the most vulnerable subcarriers would be those around $m = 35$ that are loaded with 12 bits. If the distortion due to frequency offset is not to degrade the overall SNR (about 44 dB) by more than 0.1 dB, the SDR must be > 60 dB. That is,

$$\log(35\pi\Delta) < -3.05 \qquad (8.30)$$

or $\qquad \Delta < 8 \times 10^{-6}(8 \text{ ppm}) \qquad (8.31)$

Similar calculations and simulations reported in [Zogakis and Cioffi, 1996] have shown that the permissible peak jitter on an ADSL sampling clock is approximately 2 ns.

Band Edge Timing? The scheme described above selects the best sampling time (and demodulating carrier phase) for the pilot, which is usually at the low end of the band, where the SNR is high. On short loops the usable band will extend out to $f_{samp}/2$; that is, all $(N/2 - 1)$ subcarriers will be used, and if there is any significant delay distortion in the loop, the sampling phase indicated by the pilot will not be that required for best sampling of the subcarriers near the bandedge. The question has often been asked therefore whether band-edge timing (see Section 7.5 of [Bingham, 1988]) is needed.

Because of the rapid roll-off around $f_{samp}/2$ of the transmit filter plus loop, the effects of destructive aliasing are much less severe than they are in, for example, voice-band modems, and even the worst-case sampling phase will cause only a very narrow notch at $f_{samp}/2$. This is shown in Figure 8.18 for a sixth-order low-pass filter as defined by T1.413 Issue 2 plus a 9-kft loop of 26 AWG. The difference in capacity between the best sampling time (marked by "o") and the worst (marked by "x") will depend of course on the noise and crosstalk, but it will be much less than 1%. The conclusion is that "band-edge timing" is not necessary for xDSL.

Establishing Symbol and Super-Frame Clocks. Determination of which sample of the received signal should be considered the first of the symbol is done as part of the equalizer training, so it is discussed in Section 8.4.4 and Chapter 11.

Recovery of the NTR. Figure 8.4 shows one possible method of recovering the NTR in the ATU-R, but this may be too crude; the long-term frequency of the recovered NTR will be correct, but corrections are made only once per superframe (17 ms), and for a typical frequency offset of 50 ppm the wander and the subsequent correction will be 85 ns; whether this is acceptable for use at the RT must be left as *unfinished business*.

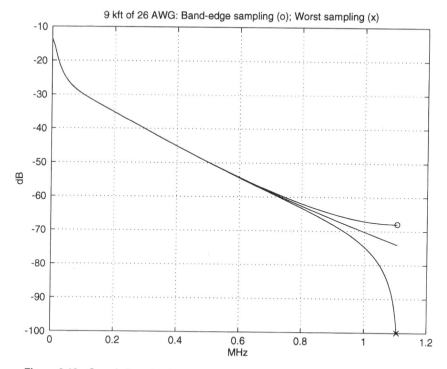

9 kft of 26 AWG: Band-edge sampling (o); Worst sampling (x)

Figure 8.18 Sampled received downstream signal with and without band-edge timing.

Phase Lock in the Central Receiver. Figure 8.19 shows the response in the upstream ADSL band of the eighth-order low-pass filter defined by T1.413 Issue 2 plus a 9-kft loop of 26 AWG. As for the downstream signal, the difference in capacity between the best sampling time (marked by "o") and the worst (marked by "x") is insignificant. The conclusion is that band-edge timing is not necessary for upstream ADSL either, and a simple division by 8 of the downstream 2.208 MHz without any selection of the phase is adequate.

8.4.4 Time-Domain Equalizers

Equalization for DMT is a very complicated subject that is still in its infancy. The first equalizers that were described in [Chow et al., 1993] used a tapped delay line sampled at the Nyquist rate (e.g., 2.208 MHz in an ATU-R) in an attempt to shorten the impulse response of the channel to the length of the cyclic prefix. The required equalizer is very similar to the feed forward equalizer (FFE) part of a DFE in an SCM system,[29] but the MCM problem should be slightly easier because there is no requirement that the first sample of the shortened impulse response (SIR) be the largest.[30] Let us assume a channel with impulse

---

[29] See, for example, [Falconer and Magee, 1973], [Messerschmitt, 1974], or [Bingham, 1988].
[30] Such an IR is sometimes called "causal," but that is inadequate; all real IRs are causal! We need a word that means "having no precursors."

response $h(D)$ $(h_0, h_1, \ldots, h_{nh})$, an equalizer $w(D)$ $(w_0, w_1, \ldots, w_{nw})$, and an SIR $b(D)$ $(b_0, b_1, \ldots, b_{nb})$.

Off-line Design of Equalizer. The problem can be represented generally in matrix form as

$$\mathbf{M} \cdot w = k \qquad (8.32)$$

where $\mathbf{M}$ is $(nh + nw - nb) \times nw$, and k includes a block of $(nb + 1)$ nonzero terms with the remaining $(nh + nw - nb)$ terms set to zero. The best position for the nonzero block (i.e., the delay through the equalizer) is not known in advance, and all $(nh + nw - nb + 1)$ possibilities should be tested; a simple example with $nh = 6$, $nw = 4$, $nb = 3$ and a delay of two is

$$
\begin{bmatrix}
h_0 & 0 & 0 & 0 & 0 \\
h_1 & h_0 & 0 & 0 & 0 \\
h_2 & h_1 & h_0 & 0 & 0 \\
h_3 & h_2 & h_1 & h_0 & 0 \\
h_4 & h_3 & h_2 & h_1 & h_0 \\
h_5 & h_4 & h_3 & h_2 & h_1 \\
h_6 & h_5 & h_4 & h_3 & h_2 \\
0 & h_6 & h_5 & h_4 & h_3 \\
0 & 0 & h_6 & h_5 & h_4 \\
0 & 0 & 0 & h_6 & h_5 \\
0 & 0 & 0 & 0 & h_6
\end{bmatrix}
\begin{bmatrix}
w_0 \\
w_1 \\
w_2 \\
w_3 \\
w_4
\end{bmatrix}
=
\begin{bmatrix}
0 \\
0 \\
b_0 \\
b_1 \\
b_2 \\
b_3 \\
0 \\
0 \\
0 \\
0 \\
0
\end{bmatrix}
\qquad (8.33)
$$

One would hope (optimistically? naively?) that the error would be a concave function of the delay, but early calculations at Amati showed a random variation; the reason for this was never found.

Because at this stage we know nothing about the desirable SIR, the $(nb + 1)$ terms must be considered "don't cares"; one of them—it does not matter which, because all the $(nb + 1)$ solutions with the same delay will be scaled versions of each other—must be set to unity to avoid the all-zero-tap solution, and the rows containing the other nonzero b terms should be temporarily discarded. Then a reduced matrix $\mathbf{M}'$ should be formed, as shown in (8.34) for the case of a delay of two and no leading zeros in b,

$$
\begin{bmatrix}
h_0 & 0 & 0 & 0 & 0 \\
h_1 & h_0 & 0 & 0 & 0 \\
h_2 & h_1 & h_0 & 0 & 0 \\
h_6 & h_5 & h_4 & h_3 & h_2 \\
0 & h_6 & h_5 & h_4 & h_3 \\
0 & 0 & h_6 & h_5 & h_4 \\
0 & 0 & 0 & h_6 & h_5 \\
0 & 0 & 0 & 0 & h_6
\end{bmatrix}
\begin{bmatrix}
w_0 \\
w_1 \\
w_2 \\
w_3 \\
w_4
\end{bmatrix}
=
\begin{bmatrix}
0 \\
0 \\
1 \\
0 \\
0 \\
0 \\
0 \\
0
\end{bmatrix}
\qquad (8.34)
$$

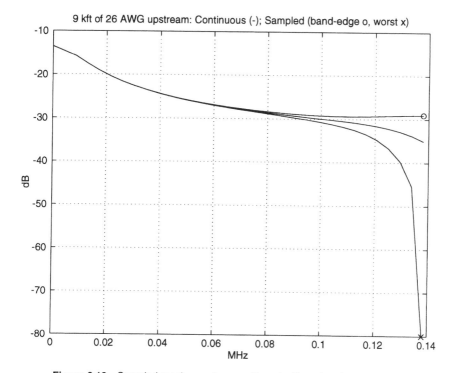

Figure 8.19 Sampled receive upstream with and without band-edge timing.

for each one of the $(nh + nw - nb + 1)$ possible delays. The mmse solution, w', for each delay is then defined by

$$\mathbf{M}'^{\mathrm{T}} \cdot \mathbf{M}' \cdot w' = \mathbf{M}'^{\mathrm{T}} \cdot k \qquad (8.35)$$

and the resulting impulse response

$$b' = \mathbf{M}' \cdot w' \qquad (8.36)$$

The Error Criterion. When the solution of (8.35) is inserted into (8.36) the right-hand-side terms that are supposed to be zero but are not represent residual distortion. The simplest and most intuitive measure of error would be the sum of the squares of these terms; this is certainly the measure that is appropriate for SCM. However, as we saw in Section 6.2, for MCM the effects of nonzero terms of the IR beyond the cyclic prefix increase linearly beyond the span of the cyclic prefix. Therefore, $\mathbf{M}'$ must be premultiplied by a row vector α to take this into account:

$$
\begin{bmatrix}
\sqrt{2}h_0 & 0 & 0 & 0 & 0 \\
h_1 & h_0 & 0 & 0 & 0 \\
h_2 & h_1 & h_0 & 0 & 0 \\
h_6 & h_5 & h_4 & h_3 & h_2 \\
0 & \sqrt{2}h_6 & \sqrt{2}h_5 & \sqrt{2}h_4 & \sqrt{2}h_3 \\
0 & 0 & \sqrt{3}h_6 & \sqrt{3}h_5 & \sqrt{3}h_4 \\
0 & 0 & 0 & \sqrt{4}h_6 & \sqrt{4}h_5 \\
0 & 0 & 0 & 0 & \sqrt{5}h_6
\end{bmatrix}
\begin{bmatrix}
w_0 \\ w_1 \\ w_2 \\ w_3 \\ w_4
\end{bmatrix}
=
\begin{bmatrix}
0 \\ 0 \\ 1 \\ 0 \\ 0 \\ 0 \\ 0 \\ 0
\end{bmatrix}
\tag{8.37}
$$

Constraining one of the right-hand-side terms to be unity does not control the other nonzero terms, and the energy in the signal pulse must be calculated as the sum of the squares of the nonzero terms. Then the best error measure for each of the delays tested is the ratio of the sum of the weighted squared errors to this signal pulse energy. For the example considered, k' has 11 terms ($k_0, \ldots, k_{10}$), and for the particular delay illustrated in (8.37) the error measure is

$$
E^2 = \frac{2k_0^2 + k_1^2 + k_6^2 + 2k_7^2 + 3k_8^2 + 4k_9^2 + 5k_{10}^2}{k_2^2 + k_3^2 + k_4^2 + k_5^2}
\tag{8.38}
$$

This design by matrix inversion is probably not practical for a modem DSP, but it should be performed off line to establish a benchmark for the equalizer designed on line.

On-line Design of Equalizer. The problem as defined in [Chow et al., 1993] is shown in Figure 8.20. The sampled transmit signal $x(D)$ is input to a channel with impulse response $h(D)$, noise $n(D)$ is added, and the sum (the received signal) is input to the equalizer $w(D)$. An SIR $b(D)$ is postulated, and the problem is to find w and b such that the error $e(D)$ is minimized *in some sense* (more anon). That is, in both time and frequency domains:

$$y(D) = h(D)x(D) + n(D) \qquad Y(\Delta) = H(\Delta)X(\Delta) + N(\Delta) \tag{8.39}$$
$$e(D) = w(D)y(D) - b(D)x(D) \qquad E(\Delta) = W(\Delta)Y(\Delta) - B(\Delta)X(\Delta) \tag{8.40}$$
$$k(D) = h(D)w(D) \qquad K(\Delta) = H(\Delta)W(\Delta) \tag{8.41}$$

All usable subcarriers are modulated with a pseudo-random sequence so as to generate a repetitive signal of just N ($=512$ for ADSL) samples (no cyclic prefix), and each block of received samples is FFTed to generate a $Y(\Delta)$. After $w(D)$ is initialized to some reasonable response[31] the basic cycle of operations is:

---

[31] The simplest is just $[\ldots,0,0,1,0,0,0,\ldots]$, but experience will probably suggest better ones.

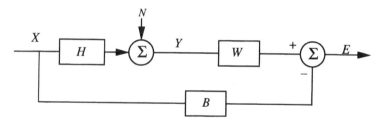

Figure 8.20   Shortening the channel IR.

1. B is updated in the frequency domain, and IFFTed to b.
2. b is windowed to $(nb + 1)$ terms, and FFTed to B.
3. W is updated in the frequency domain, and IFFTed to w.
4. w is windowed to $(nw + 1)$ terms, and FFTed to W.

The updating of B and W can be performed by either division ($B = WY/X$ and $W = BX/Y$) or LMS adaptation. This results in four possible combined strategies, but the one that has proved most useful is division for B and LMS updating for W. There are three reasons for this—two practical and one intuitive:

1. Division is a tedious operation in DSP, but since X is a known vector, its inverse can be calculated off-line and stored, and the division performed by a multiplication.
2. It is easier to choose an initial value for W than for B.
3. It is W that we are seeking, and B, in a sense, follows it.

The LMS update of W [from the kth to the $(k + 1)$th value] is defined by

$$E_k = B_{w,k}X - W_{w,k}Y_k \tag{8.42}$$
$$W_{u,k+1} = W_{w,k} + \mu EY^* \tag{8.43}$$

where the suffices w and u indicate windowed and unwindowed; that is, (8.43) shows that a new W, which will be the FFT of a w that has more than $(nw + 1)$ terms, is calculated by update of an old W, which was the FFT of a windowed w.

The windowing operations on w and b are combined with a timing adjustment. At each stage the block of $(nw + 1)$ or $(nb + 1)$ terms that has the largest total energy should be selected, and the terms outside those blocks zeroed. If the position of either of these blocks has shifted since the last iteration, the index of the other must be adjusted accordingly.[32]

---

[32] Beware, this is messy!

Problems. This method has three serious problems:

1. The iteration does not always converge or may converge to a local, nonglobal minimum; the choice of an initial b and step size μ are critical.
2. LMS updating of W in the frequency domain is equivalent to an equal weighting of the error terms of the resultant IR. As we have seen, a linearly increasing weighting is a better measure of the distortion. Perhaps the frequency-domain multiplications of (8.36) could be replaced by convolutions, and the time-domain error function, e, weighted appropriately.
3. There is no control over the dynamic range of W, and excessive noise enhancement, as described in Section 6.3, may result.

In addition, there is the much more fundamental problem of equalization in the time domain, which, as shown in Section 6.2.3, ignores the effects of distortion on the separate, differently loaded subcarriers. Some attempts to solve these and other equalizer problems are discussed in Chapter 11.

8.4.5   FFT

Techniques for implementation of an FFT are described in Appendix C.

8.4.6   Frequency-Domain Equalizer

If the time-domain equalizer has done its job, and the resultant channel response is an SIR with no more than $(\nu + 1)$ terms, then all the modulated subcarriers delivered to the FFT will be attenuated and rotated by different amounts, but they will not affect each other (i.e., they will be orthogonal). The detection and decoding are greatly simplified if all subcarriers have the same amplitude and phase, so the frequency-domain equalizer must correct for the attenuation and rotation. That is, it multiplies the FFT output by a diagonal matrix (i.e., with one complex multiplication per subcarrier) whose elements are the inverse of the transform of the SIR.

8.4.7   Trellis Decoder (Viterbi Decoder)

The differences between trellis encoding used for SCM and MCM are discussed in Section 8.2.7. With these in mind, the necessary changes to a conventional decoder (see, e.g., [Bingham, 1988] or [Kurzweil, 1999]) should be apparent. One important point that should be noted is that for a trellis code that operates over time (i.e., SCM) the desideratum is that the noise be uncorrelated in time; the front-end filter should therefore ideally be a nose-whitening filter. For the MCM encoder that operates across frequency the dual desideratum is that the noise should be uncorrelated in frequency; if the noise is dominated by crosstalk from other MCM signals, this is more or less guaranteed; I am not sure how well this is met if the crosstalkers are SCM.

NOTE: If trellis coding is not used, this stage is a simple memoryless constellation decoder.

8.4.8   De-interleaver

This is an exact reverse of the interleaver discussed in Section 8.2.4; it can be understood by simply replacing "write address" wherever it occurs with "read address," and vice versa.

8.4.9   Reed–Solomon Decoder

This is a conventional decoder about which I know nothing. Readers are referred to [Berlekamp, 1980] and [Lin and Costello, 1983].

8.4.10   Descrambler

This is a conventional self-synchronizing descrambler (see [Bingham, 1988]), which (unnecessarily) triples the bit error rate—and either doubles or triples the byte error rate—out of the R-S decoder. The argument has been made that the tripling has no significant effect on any application, but I do not know how well that has been substantiated.

8.5   ALGORITHMS (PART TRANSMITTER AND PART RECEIVER)

8.5.1   Channel Measurement

The main requirements for channel measurement are common to all DMT modems; they are discussed in Section 5.2. The specifics for ADSL are defined in T1.413 and G.992. There may be some extra measurements required for *bit rate assurance*; these are discussed in Section 8.5.5.

8.5.2   Bit Loading

ADSL is strictly PSD limited, so transceivers can use either of the allocation methods in Section 5.3.

8.5.3   Bit Rate Maintenance (Bit Swap)

When the SNR of a channel changes, the SNR values of individual subchannels will usually change differently. The most likely cause of change is another user in the same binder group either turning on or off and changing the level of crosstalk at various frequencies across the band.[33] The change in aggregate SNR may be small, but the change on a few subchannels may be significant. Strictly speaking, improvements and deteriorations should be treated equally, but all the attention has been focused on the situation when the SNRs on a few subchannels decrease.

If the bit loading was done correctly, then at the beginning of a session the SNRs (and hence the probabilities of error) on all subchannels should be the same. Therefore, a receiver continually monitors the SNRs by noting the slicing errors from every decision, and averaging their squared values over, typically, a few hundred symbols. For "symmetrical" operation a bit swap should be initiated if the difference between the highest and lowest SNR values exceeds some threshold (typically, 3 dB); if it is desired to protect only against decreases in SNR, a swap should be initiated only if the lowest SNR falls below a threshold (typically, 3 dB below that set during loading).

A bit swap is exactly that. The receiver sends instructions—via an overhead channel—to the transmitter to decrease by one the number of bits encoded onto one tone and simultaneously increase the number encoded onto another. If receiver and transmitter keep track of superframe numbering, and make the change at the beginning of the same superframe, this will result in no errors. It is interesting to note (see the next section on DRA) that as defined in T1.413 and G.992, decrease and increase are separate instructions; it is possible to do one without the other even though this would not be a "swap".

Field experience has shown that bit swap is very useful; individual subchannel SNRs do change significantly. It is important to note, however, that this will not be observed in tests with crosstalk simulators, which generate crosstalk that varies smoothly with frequency, so that all subchannels improve or deteriorate together.[34]

8.5.4 Dynamic Rate Adaptation

If the SNRs change enough that the overall margin decreases by more than about 3 dB, bit *swap* will no longer suffice; bit *drop* will be necessary. In the PMD layer this *could* be done without any loss of data by sending decrease instructions that are "postdated" with a later symbol number: that is, not to be implemented until a full set has been received. Unfortunately, T1.413 and G.992.1 do not do it this way; they define a sequence that results in breaking the link for several tens of milliseconds.

The big problem with dynamic rate adaptation is not in the PMD layer, but as discussed by Alan Weissberger in Chapter 2, in the TC layer and above. Rate reduction means breaking a *traffic contract*. Since constant bit rate (CBR)

---

[33] Recall that individual pair-to-pair crosstalk transfer functions—particularly those between close pairs—can vary rapidly with frequency.
[34] This may explain why many people have claimed that bit swap is not needed—they were basing their conclusions on lab tests. The fact that bit swap is patented [Hunt and Chow, 1995] may also, of course, have had an influence!

contracts are more important than available bit rate (ABR) contracts, it would seem that only the latter should be broken. How this might be done is discussed in the next section.

8.5.5   Unfinished Business: Bit Rate Assurance

It has been said several times in this book that one of the biggest challenges for the wide deployment and usage of ADSL is how to take full advantage of the widely varying data rates available. A customer located 20 kft from a CO signing on at 4.00 A.M. might be able to use 3 Mbit/s downstream if his applications were such that he could accept a big drop in data rate as soon as his neighbors sign on. His initial contract could therefore be for a (more or less assured) CBR that is appropriate to his distance from the CO and how many other *potential* ADSL users there are in his binder group, plus ABR up to what the present SNRs will allow. How much information would have to flow between system management and the modem to achieve this remains to be determined.

9

COEXISTENCE OF ADSL WITH OTHER SERVICES

The "other services" to be considered here are voiceband and the three types of BRI (two echo-canceled, one TDD) defined in G.961. Coexistence in the first three cases (Sections 9.1 and 9.2) means on the same UTP, with special problems associated with the filtering needed to separate the services. In the fourth case (Section 9.3) it means only in the same binder group; the problem is the now familiar NEXT but at such a level that it requires special techniques to make ADSL viable.

9.1 COEXISTENCE WITH VOICE-BAND SERVICES

These have been traditionally called POTS (plain old telephone services), and it has been said that they use the "voice" band. Many of the services are no longer voice, and some of them (e.g., V.90 modems) are neither plain nor old,[1] but nevertheless, we will use the old terminology. Originally, the band extended only to 3.4 kHz, but now with the advent of high-speed modems (paricularly V.90) it extends up to 4 kHz; it would be useful to have the ADSL band extend down as close to the 4 kHz as possible, but 25 to 30 kHz is a practical lower limit.

ADSL and POTS are separated by a bidirectional low-pass/high-pass filter pair[2] as shown in Figure 9.1. The input impedance of each filter must go high out of band (so as not to load the other filter), so they must begin with a series inductor/capacitor as shown. T1.413 and G.992 consider the high-pass filter to be part of the ATU, and the term *splitterless* that is used to describe ADSL lite really means "low-pass filterless." It is more conducive to an understanding of the overall problem, however, to consider the three-port network as one separate device.

If the POTS device is a modem that presents a linear, resistive impedance to the line, the pair needs to perform just three tasks:

---

[1] [Starr et al., 1999] said that P(ositiveley)A(mazing)N(ew)S(ervices) had been suggested as a name, but I think Tom made that up!

[2] It is sometimes thought that these filters must be passive, but active low-pass filters as described in Section 9.1.4 are essential in some telephone systems.

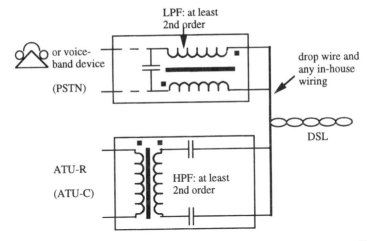

Figure 9.1 POTS splitter as a three-port network with low-pass and high-pass filters.

1. The low-pass protects the ADSL signal from the high-frequency transients (particularly off-hook and ring trip) associated with POTS signaling.
2. The high-pass isolates the voice-band device from the relatively low input impedance (typically, 100 Ω) of the ATU.
3. The high-pass, together with any internal filters, protects the voice-band device from the out-of-band ADSL signals and the ADSL signal from the low-frequency steady-state POTS signaling (dialing and ringing).

If, however, the voice-band device is a telephone, the pair may have two more tasks:

4. The low-pass isolates the ATU from the capacitive input impedance of some telephone handsets, which would otherwise provide a low shunt impedance to the high-frequency ADSL signals.
5. The low-pass isolates the nonlinear input impedance of some telephones from the high-frequency ADSL signals, which might otherwise be intermodulated down into the voice band.

Furthermore, the low-pass must pass the loop current needed for all POTS operations,[3] and should, ideally, not degrade the voice-band performance. Task 1 for the low-pass is the most difficult; performing it would almost certainly ensure performing tasks 4 and 5. For ADSL lite, however, task 1 is greatly relaxed; then task 5 becomes important also.

---

[3] Traditionally specified as a maximum of 100 mA but that can occur only on a short loop, on which the voice quality would otherwise be high.

Location of the Splitter. A "full" splitter that performs all of the foregoing tasks is, as we shall see, a fairly complex device. It was originally intended that it should be placed at the service entrance [a.k.a. network termination (NT) and network interface device (NID)], and that separate and separated[4] in-house wiring should be used for POTS and ADSL services. This was deemed, albeit grudgingly, to be acceptable for a sophisticated, very high-speed ADSL service (up to 6 + Mbits/s) for which all equipment was to be owned and maintained by the LEC, but it was later realized that it is unacceptable for a more plebian service (ADSL lite) for which the ATU-R is owned and installed by the customer. The simplification and relocation of the low-pass filter are discussed in Section 9.1.6.

9.1.1 Transient Protection for the ATU

The worst transient is ring trip, seen at the CO. When the telephone recognizes the ring signal it goes off-hook and draws loop current; this is detected by the circuitry at the CO, which then switches from the ringing signal[5] to the -48-V battery, generating a quasi step function. If the telephone does not have special circuitry to trip the ring at a zero crossing, it may occasionally trip it near a positive peak, and the transient at the CO will be a nearly instantaneous transition of almost 150 V: current-limited in some circuits, but not in all.

Much of the expertise that was brought to the discussions of this transient during the early work on T1.413 has been lost, and only the conclusion has survived. This was that in order to prevent ADSL errors, the low-pass filter should provide between 70 and 80 dB of attenuation at 30 kHz. In Issue 2 of T1.413, with a better understanding of the operation of the FEC function and with more appreciation of the difficulty of designing and building the filters, this requirement was relaxed to 65 dB [including the 8 dB or so of attenuation that comes from the mismatch of the 600-Ω source and approximately 150-Ω (at 30 kHz) loop]. Ring trip seen at the RT is attenuated somewhat, but the 65-dB attenuation requirement was maintained.

NOTE: This requirement for 65 dB at RT assumes that the downstream band extends down to 30 kHz (i.e., EC is used). At about 150 kHz, which is the low end of the downstream band for an FDD system, all POTS transients would be further attenuated, and furthermore, the low-pass filter could be designed to have considerably more attenuation.[6] I have not seen a realistic assessment of the low-pass attenuation (different at CO and RT) needed for an FDD system.

---

[4] Physical separation of the pairs is necessary to prevent the POTS signaling transients from NEXTing into the ATU-R.

[5] 20 Hz in the United States, only slightly (insignificantly for our present purposes) different elsewhere in the world.

[6] The second-order minifilters discussed in Section 9.1.6 would have 28 dB more.

9.1.2 Isolating the Voice Band from the (Low) Input Impedance of the ATU

This is a very difficult requirement to quantify. As noted above, T1.413 and G.992 consider the high-pass filter to be part of the ATU and the low-pass filter to be a separate device that must be specified independently. Therefore, they define the input impedance of the high-pass/ATU that must be assumed when designing the low-pass filter. This was probably the only practical way to deal with the problem, but the purpose of the isolation is to help maintain the quality of the voice-band service, and this can be done better if the two filters are designed together. The problems are considered all together in Section 9.1.5.

9.1.3 Maintaining Voice-Band Quality

The three characteristics of voice-band performance that should, ideally, be maintained are noise level, end-to-end response, and return loss. If the loop was loaded before the ADSL service was installed, the question for response and return loss is: What should be maintained? The rules for adding loading coils—and therefore the quality of the "pre-ADSL" service to which the customer is accustomed—seem to vary from LEC to LEC: Some say any loop over 12 kft may be loaded; others say only loops over 18 kft are loaded.

Noise Level. The attenuation of the lower sidelobes of the upstream signal and the high-pass filter should together, ideally, keep all ADSL signals out of the voice band. If, however, the input impedance of the telephone is nonlinear, components above 30 kHz can be intermodulated into the voice band; task 4 in Section 9.1 becomes important. If there is no low-pass filter, the upstream transmit level must be reduced to less than $-46\,dBm/Hz$ (a reduction of at least 8 dB) to keep the intermodulation products to a tolerable level.

End-to-End Response. Figure 9.2 shows the response of 13.5 kft of 26 AWG (one of the test loops defined in T1.413) both loaded and unloaded. It is clear that removing the loading coils greatly degrades the response. It is impractical—perhaps even theoretically impossible—to restore the response to that with loading coils, but it is possible to reduce the roll-off slightly by using minifilters, as shown by the third plot (marked with asterisks) and discussed in Section 9.1.6.

Return Loss. In the United States, three types of return losses are defined for voice use:

1. *Singing return loss (SRL) low:* the minimum return loss in the band 300 to 500 Hz
2. *Echo return loss (ERL):* the "average" return loss (obtained by comparing total transmitted and reflected powers) in the band 500 to 2500 Hz
3. *SRL high:* the minimum return loss in the band 2500 to 3400 Hz

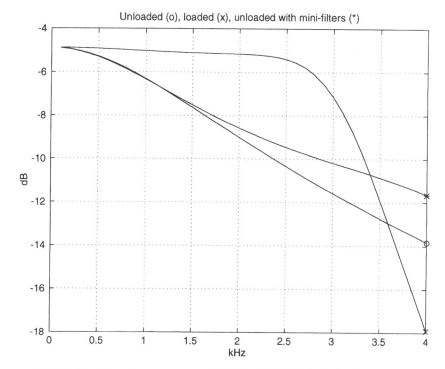

Figure 9.2   Voice-band responses of 13.5 kft of 26 AWG: loaded and unloaded.

ITU Recommendations G.122 and G.131,[7] however, define the band for SRL high all the way out to 4.0 kHz.

The problem of controlling these is greatly complicated in the United States by two factors:

1. Even though, after the loading coils have been removed, a loop is almost symmetrical in the voice band,[8] different terminating impedances are used at the CO and RT: 900 Ω in series with 2.16 μF, and 600 Ω, respectively. This makes the input impedances seen at the two ends different and requires different filters at CO and RT.

2. The terminating impedances are resistive (or almost so) instead of a first-order bilinear RC (RRC) approximation to the characteristic impedance that is used in the United Kingdom and many other European countries. This means that the input impedances vary considerably with loop length, and all designs have to be compromises.

At both ends of the loop, attempts are made to match the loop impedance and balance the 4W/2W network. In a telephone this reduces the sidetone from

---

[7] This is a very old reference, which I got from [Freeman, 1981]; it may be out of date.
[8] Gauge changes and bridge taps have very little effect in the voice band.

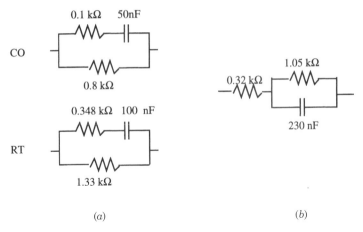

(a)   (b)

Figure 9.3   Compromise RRC impedances: (a) U.S. CO and RT; (b) U.K.

microphone to speaker; in a modem it lightens the load of the echo canceler, and—probably most important—at the CO it reduces the delayed echo that the far end experiences. The compromise matching impedances are RRC as shown in Figure 9.3.

The return losses at the CO relative to the RRC impedance of Figure 9.3(a) for the loaded and unloaded 13.5-kft loops are shown in Table 9.1, together with the requirements in Issue 1 of T1.413 *with the splitter installed.* It can be seen that the loaded loop does not completely meet the requirements *even without the ADSL service.*

Figure 9.4 may help to explain what is happening; it shows the loci with frequency of the RRC impedance and the input impedance at the CO of both loops. It can be seen that the impedance with loading coils is very different from the RRC impedance. The return loss situation is thus the reverse of the response one: it is poor with loading coils and good when they are removed.

The situation deteriorates again, however, when LC low-pass filters are inserted. The requirement of 65 dB attenuation at 30 kHz can be met with a fourth-order elliptic function filter (C041014 in [Zverev, 1967]) with a cutoff frequency of 6.8 kHz. If terminated in a 600-Ω resistance the locus of the input impedance would be a small quasi-ellipse, and the minimum return loss would be 20 dB. If this filter is terminated by the loop, however, then, as pointed out in

TABLE 9.1   Voice-Band Return Losses for Loaded and Unloaded Loops

	SRL Low (dB)	ERL (dB)	SRL to 3.4 kHz (dB)	SRL to 4.0 kHz (dB)
13.5 kft 26 AWG unloaded	8.9	12.1	16.4	16.4
13.5 kft 26 AWG loaded	8.9	10.7	2.8	2.8
T1.413 requirements	5	8	5	n/a

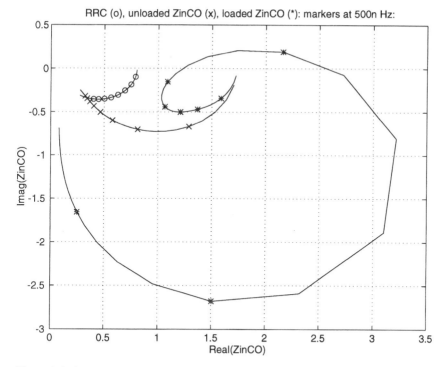

Figure 9.4 Input impedance at CO of 13.5-kft loops with impedance of compromise RRC.

[Cook, 1994], the locus expands greatly, and the return loss relative to the compromise RRC impedance falls to nearly 0 dB at the edge of the passband. Figure 9.5 shows the input impedances of a 9-kft 26-AWG loop (short enough never to have needed loading) with no filter and the above fourth-order filter. Table 9.1 also shows the return losses for this loop: with and without filters.

As a way of explaining this effect, Cook suggested that the filter behaves in its passband like a lossless transmission line. At dc and all other frequencies at which $\gamma l = n\pi$, the filter is transparent and $Z_{\text{in}} = Z_{\text{loop}}$, but at frequencies at which $\gamma l = (n + 1/2)\pi$ the filter has the well-known effect of inverting its terminating impedance (i.e., $Z_{\text{in}} = Z_0^2 / Z_L$, and for example, an open circuit looks like a short circuit). This means that at the latter frequencies the negative imaginary part of the "RC" impedance is transformed to positive. Thus the input impedance is RLC and oscillates between these two extremes; the number of the oscillations and the frequencies at which they occur depend on the order of the filter, the cutoff frequency, and the sharpness of the cutoff. Some improvement in the return loss can be achieved by increasing the order of the filter[9] and pushing the cutoff frequency much higher, but my own calculations and published results

---

[9] A sixth-order quasi-antimetric filter is common in the Unites States.

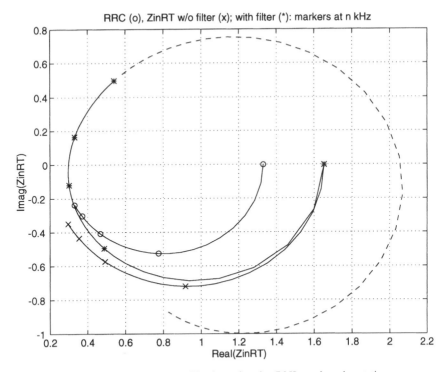

Note: The plot with the filter is continued to 7 kHz to show the rotation
throughout the passband; the others are curtailed at 4 kHz.

Figure 9.5   ZinRT of 9-kft 26 AWG with and without fourth-order filter plus compromise RRC.

[Cook, 1994] and [Hohhof, 1994] suggest that there is no standard "textbook" filter that meets all requirements.

9.1.4   One Solution to the Impedance Problem: Generalized Immittance Converters

The most elegant solution to the impedance problem as it exists in the United Kingdom was described in [Cook, 1994] and [Cook and Sheppard, 1995]. Loading coils are not used in the United Kingdom, so all loop input impedances are RC, and both the line-driving/terminating impedance and the balancing impedance in the 4W/2W network are compromise RRC, such as are shown in Figure 9.3(b): trying to match the characteristic impedance of the loop. Many of the telephones are active and require a return loss of about 18 dB or better.

A generalized immittance converter (GIC) is a reciprocal active two-port defined by

$$Z_{in1} = F_1(p)Z_{load2} \quad \text{and} \quad Z_{in2} = \frac{1}{F_1(p)}Z_{load1} \tag{9.1}$$

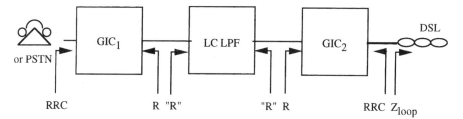

RRC   R   "R"   "R"   R   RRC Z_{loop}

Note: "R" is the "almost resistive" input impedance of an
LC filter with a "flat" passband terminated in R_{LC}.

Figure 9.6 Impedance transformations using GICs.

and the solution uses a mirror-image pair of them at each end arranged as shown in Figure 9.6. The requirements for both GICs are

$$Z_{in1} = \frac{R_\infty(p + p_1)}{p + p_2} \quad \text{and} \quad Z_{in2} = R_{LC} \tag{9.2}$$

In-line GIC. One implementation uses the two-amplifier circuit described in [Antoniou, 1969] and [Fliege, 1973].[10] This is shown in its most basic form in Figure 9.7. The input, output, and "bc" nodes are connected to the inputs of the two op-amps so as to keep the three nodal voltages V_1, V_2, and V_{bc} equal. Then the impedances seen at the two ports are

$$Z_{in1} = \frac{Z_a Z_c}{Z_b Z_d} Z_{L2} \quad \text{and reciprocally} \quad Z_{in2} = \frac{Z_b Z_d}{Z_a Z_c} Z_{L1} \tag{9.3}$$

If the amplifiers are considered to be ideal, there are (theoretically) many ways of connecting the nodes to the op-amp inputs and assigning the pole and zero of (9.2) to $Z_{1,2,3,\text{and}4}$, but the requirement of high-frequency stability with real

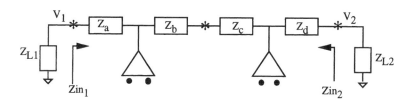

Note: Theoretically, the three nodes ✳ may be connected to inputs of the
four op-amps● in several different ways; practically, only a few are
stable and only a very few are usable.

Figure 9.7 Two-amplifier implementation of a GIC.

---

[10] Variations of this circuit are also used to implement biquadratic filter sections, gyrators, and frequency-dependent negative resistors.

amplifiers reduces the number of workable combinations to a very few; unfortunately, details of the preferred circuits are proprietary and unpublished.

Off-line GIC. The Fliege circuit maintains $V_1 = V_2$ and scales the currents; the circuit described in [Cook, 1994] on the other hand, maintains $i_i = i_2$ and scales the voltages. It does this by adding a voltage in series between the ports:

$$V_2 = V_1 + V_1 \left[\frac{R_\infty(p + p_1)}{R_{LC}(p + p_2)} - 1 \right] \tag{9.4}$$

It is desirable that the added part in brackets be zero at infinite frequency (i.e., that the active circuitry be low-pass). Therefore, R_{LC} is set $= R_\infty$, and

$$V_2 = V_1 + V_1 \frac{p_1 - p_2}{p + p_2} \tag{9.5}$$

Implementation. V_1 can be detected from the balanced port 1 by a high-impedance (so as not to load the port) difference circuit, multiplied by the first-order low-pass transfer function in (9.5), and then added in to the series voltage via a third winding on a balanced (i.e., with zero differential-mode inductance) "summing" inductor. This is shown in Figure 9.8, which is a simplified version of Figure 8 of [Cook, 1994].

Both low- and high-pass filters must be balanced about ground. Figure 9.8 shows how each series inductor can be economically implemented as a balanced pair wound on the same core. The shunt capacitors can be implemented as one (C_1) between the balanced arms or as two double-size capacitors ($2C_2$ and $2C_2$) to "ground". The relative merits of these two depends on the need for common-mode filtering and on the availability of a good ground; analog designers do not all agree on which is better.

The advantage of the "off-line" circuit is its robustness; if power is lost, the output impedance of the driver op-amps goes high, the third winding on the

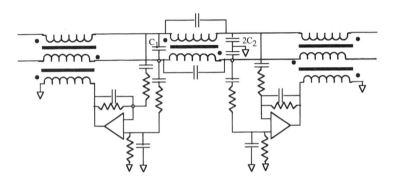

Figure 9.8 Balanced implementation of low-pass and "off-line" GIC.

summing inductor is open circuited, and the GIC has no effect; the voice-band signal is maintained, albeit with degraded impedances.

Use of the GIC Approach in the United States. It was suggested in [Cook, 1994] that U.S. needs (with quasi-resistive terminations) could be met with just one GIC on the line side of the filter at each end. This, however, would only partially solve the problem: the impedance seen from the telephone or the CO would now be resistive instead of the expected RRC. The return loss of this relative to the RRC compromise would be better than it would be with no GIC at all, but whether it would be good enough requires careful study.

9.1.5 A Partial Solution: Custom Design by Optimization

As noted in Section 9.1.3, it seems fairly certain that no pair of conventional (designed to work between resistive impedances) LC filters can meet both the attenuation and the return loss requirements. However, using a pair of these filters as a starting point, and iterating on their components simultaneously to minimize a weighted sum of errors (low-pass passband response and return loss and stopband response, and high-pass response), can result in a significant improvement.

NOTE: The weightings will almost certainly have to be adjusted as the iteration progresses, so as to keep each of the performance parameters reasonably within bounds.

Unfortunately—for the book—programs to do this iterative design are proprietary. All I can say (as an incentive!) is that echo return losses of 12 dB can be obtained with most loops using a sixth-order low-pass. This does, however, require simultaneous adaptation of both low-pass and high-pass; it is doubtful that 12 dB can be achieved when the low-pass is designed as a separate unit and a fixed compromise input impedance is used to emulate the high-pass.

The resulting passband responses are far from equal ripple: Loss and mismatch at low frequencies seem to push out the frequencies at which $\gamma l = (n + \frac{1}{2})\pi$ and to "postpone" the input impedance becoming inductive. The return losses are not as good as could be obtained with an unloaded loop and no filters, but they are considerably better than could be obtained (in the United States at least) with loaded loops. The passband response is not as good as with a loaded loop, but it is probably better than with an unloaded loop without a filter.

9.1.6 Simplified (Dispersed and Proliferated) Low-Pass Filters[11]

One way of avoiding the dreaded "truck roll" would be for a customer to buy (full-size) low-pass filters and install them at every POTS device; the transients

---

[11] These have been called *in-line* filters, but this is a misleading name; they are not in the line (loop?), they are associated with every POTS device. *Minifilters* is a better name.

would be filtered at the source, and all POTS-associated signals on the house wiring would be confined to the voice band. These filters, however, have a minimum of three inductors each and are bulky and expensive.

Early expectations for ADSL lite were that these filters could be eliminated altogether, and the only problem would be the short bursts of errors (perhaps correctable, but if not, then probably tolerable) caused by POTS signaling. It was soon realized, however, that tasks 4 and 5 in Section 9.1 are ongoing; with no low-pass filters some telephones would receive an intolerable level of noise whenever the ATU-R was transmitting,[12] and some would almost short out the ADSL loop at high frequencies whenever they were off hook.

NOTE: The difference between protecting against transients (task 1) and against steady-state effects (tasks 4 and 5) is important for the TC layer. For many applications a burst of errors upon ring trip would be tolerable (as long as neither layer, PMD or TC, lost "sync"), but a subsequent change of state that resulted in a significantly lower capacity would be very difficult to deal with.

An attractive compromise between full-size filter and no filter is a simple second-order (one inductor, one capacitor) *minifilter* that performs tasks 4 and 5 and probably protects against most signaling transients from most phones.[13] Figure 9.9 shows a typical house wiring with minifilters plugged into every RJ11 jack.

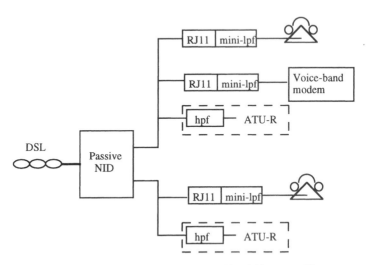

Figure 9.9 Typical in-premises wiring with mini low-pass filters.

---

[12] There was a suggestion to reduce the level of upstream transmission, but this would reduce the range when the upstream was subjected to alien NEXT.

[13] "Probably" and "most" are very imprecise words, but they are the best I can do at this time. Tests are needed with many different loops and phones.

NOTE: These filters must be balanced, and because there will not usually be a ground available at the jack, the shunt capacitor must be a single one between the balanced arms; no common-mode filtering can be performed.

A typical filter would have a nominal cutoff frequency of about 4 kHz (achieving about 34 dB of transient protection at 30 kHz), with $L \approx 28$ mH and $C \approx 50$ nF. The magnitude of the input impedance from the loop side with a telephone-induced shunt capacitor load ≈ 5 kΩ at 30 kHz, so the load on the nominally 100-Ω loop is negligible (task 4).

Input Impedance and Return Loss. Even such a minimal filter still rotates the RRC loop impedance slightly, and the return losses above about 2 kHz are between those for full-size and no filters. This is shown in Figure 9.10 for a 9-kft 26 AWG loop, and the return losses are shown in Table 9.2

Response. A bonus feature of minifilters is that they improve the voice-band response slightly. The series inductors act somewhat like loading coils and partially cancel the shunt capacitance of the loop. The third curve in Figure 9.2 shows the response of the 13.5-kft 26 AWG loop with two minifilters added. This loop may have been loaded before ADSL installation, and the customer may

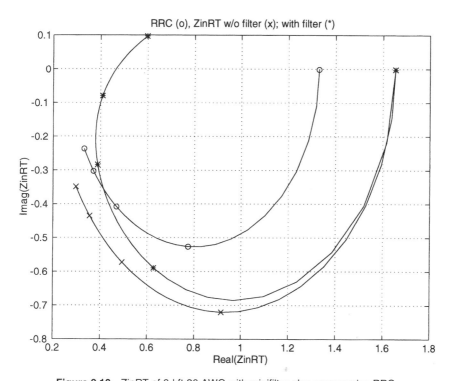

Figure 9.10   ZinRT of 9-kft 26 AWG with minifilter plus compromise RRC.

TABLE 9.2   Voice-Band Return Losses with and Without Filters

	SRL Low (dB)	ERL (dB)	SRL to 3.4 kHz (dB)	SRL to 4.0 kHz (dB)
9 kft 26 AWG				
Without filter	19.1	18.6	17.6	17.3
With full filter	17.7	10.6	2.4	1.5
With minifilter	19.7	18.4	9.6	6.9

have become accustomed to a flat response, so even a slight flattening may be welcome.

Minifilters at the CO. All the emphasis has previously been on simplifying the remote filters, and the tacit assumption seems to have been that the CO would still use a full-sized filter. In view of the better return loss achieved with the minifilters, a strong argument could be made for using them at both ends. The high-frequency components of the signaling transients imposed on the upstream signal would, of course, be increased; whether they are tolerable may depend on whether interleaving is used for upstream (see Section 2.4.3).

Minifilters with Active Telephones and/or Complex Terminating Impedances. Table 9.3 shows the four return losses (at either CO or RT because the complete circuit is now symmetrical) for 9-kft and 13.5-kft 26-AWG loops[14] with terminating and reference impedances as shown in Figure 9.4(*b*): with no filter, a second-order filter, and a fourth-order filter. It can be seen that for the average-length loop the minifilter actually *improves* the return loss over most of the passband; for the long loop it improves the ERL but significantly worsens the SRL high. It remains to be seen whether the SRL high can be either improved or

TABLE 9.3   Return Losses with Complex Terminations: with and Without Filters

	SRL Low (dB)	ERL (dB)	SRL to 3.4 kHz (dB)	SRL to 4.0 kHz (dB)
9 kft 26 AWG				
Without filter	15.2 (note)	18.6	15.1	14.4
With full filter	15.1	15.5	3.2	1.7
With mini-filter	15.7	22.5	18.0	8.9
13.5 kft 26 AWG				
Without filter	13.0	17.9	15.3	14.7
With full filter	13.1	12.5	2.9	2.6
With minifilter	13.4	21.9	9.1	5.9

Note: The values without any filters are probably slightly too low because the compromise impedance is for UK cable, and the RLGC parameters used were for U.S. cable!

[14] I know! Loops in the United Kingdom are measured in kilometers and they are metric gauge, but since the point here is comparison with what we have discussed previously, it is more informative to stay with U.S. units.

tolerated, and whether minifilters can be used instead of full-size filters and GICs.

9.2 G.992 ANNEX B: COEXISTENCE WITH ECHO-CANCELED ISDN

In many countries with a large installed base of ISDN modems, it is necessary that ADSL operate in the frequency band above ISDN. The transmit signals as defined in T1.601 and G.961 Annexes I and II (2B1Q and 4B3T) have nominal -3-dB points at 80 and 120 kHz, respectively, and are only very lightly filtered, so if the crossover frequency between ISDN and ADSL had been continuously variable, it would have been hard to get agreement on the permissible low end of the ADSL. DMT as defined in T1.413 had, however, already established bands of 138 kHz (comprising 32 subcarriers): band 1 for upstream, bands 2 through 8 for FDDed downstream. It was very easy (and probably nearly optimal) to reserve the band 1 for ISDN, band 2 for upstream, and bands 3 through 8 for downstream.[15]

The most straightforward way of encoding the upstream data into band 2 is to constellation-encode just as for a T1.413 or G.992.1 Annex I modem, and then modulate onto tones 33 through 63 using a 128-pt IFFT with tones 1 through 32 zeroed. However, before the issue of ADSL over ISDN was raised, at least one manufacturer had firmly embedded a 64-pt FFT in silicon, and their only feasible way of moving the data up to band 2 (and down again in the receiver via the reverse process) was to modulate it to the lower sideband of an IF carrier at 276 kHz (very easily generated). This has the effects of interchanging the sidebands of each narrow QAM signal and thereby changing the constellation encoding rules. ETSI TM 6 and ITU SG 15 are currently working on the problem of ensuring compatibility between a transmitter that uses one method and a receiver that uses the other.

9.3 G.992 ANNEX C: COEXISTENCE WITH TDD ISDN

TDD[16] ISDN as defined in Annex III of G.961,which is used primarily in Japan, uses a bipolar [a.k.a. alternate mark inversion or (AMI)] pulse at a symbol rate of 320 kbaud with very little filtering. The transmit PSD is shown in Figure 9.11 together with those of HDSL and downstream ADSL for comparison. The total data rate, which is slightly less than 320 kbit/s because of a guard period to allow for propagation delay, is shared equally between upstream and downstream using TDD with a superframe period of 2.5 ms (rate $= 400$ Hz). The XT coupling

---

[15] Kindred NEXT would make the band above 138 kHz unusable for full-duplex transmission, so there can be no EC option for ADSL above ISDN.

[16] This was originally called time compression multiplex (TCM), but for ADSL TCM means trellis coded modulation. The ITU has, unfortunately, perpetuated the use of the confusing TCM, but I will stick with TDD.

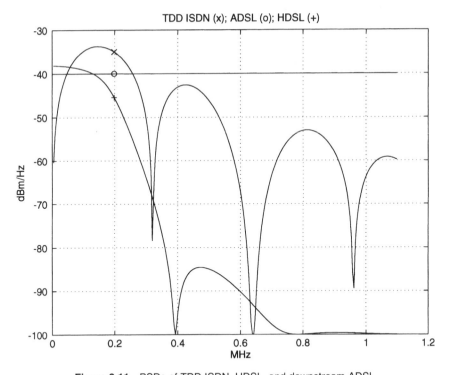

Figure 9.11 PSDs of TDD ISDN, HDSL, and downstream ADSL.

coefficients are given in Table 3.6; they are higher than anything encountered previously.

ADSL service in the same cable as TDD ISDN is very different—both in crosstalk conditions and in the duplexing method that those conditions dictate—from any considered previously. The main problem is the time-varying nature of the crosstalk; when the downstream ADSL is going "with the TDD flow" it incurs alien FEXT, which is milder than kindred FEXT, but when it is going against the flow it incurs alien NEXT, which may be very severe. Therefore, ADSL may use at least two different bit loadings: heavy for FEXT and light for NEXT. If the ADSL and TDD clocks are not frequency-locked, the ADSL symbols will precess slowly through a TDD superframe and incur changing crosstalk. A continually changing bit loading to match this would be very inconvenient, so either the clocks must be locked or the light loading must be used whenever there is *any* NEXT (protecting the error rate, but inefficient).

9.3.1 Synchronizing TDD ISDN and ADSL

Synchronization is very easy—the 400-Hz clock is available at the CO for all ATU-Cs to use—but a crucial question is: What should be synchronized? The *data* symbol rate for ADSL is 4 kBaud, exactly 10 times the frame rate for TDD ISDN. The on-line symbol rate as defined by T1.413 and G.992 Annex A, however, is $(69/68) \times 4\,\text{kBaud}$, to allow for the insertion of one sync symbol after every 68 data symbols. Therefore, *if* the symbol length (544 samples) of Annex A is preserved, the on-line symbols will precess slowly with respect to the TDD ISDN, and line up only once every 690 symbols: the synchronization must be 10 ADSL superframes (690 symbols) = 69 TDD frames.

Unfortunately, the 544-sample symbol was preserved[17] in Annex C, and the result is a complicated and inefficient framing structure. It is complicated because it must keep track of the crosstalk/loading status of each of the 690 symbols of a TDD *hyperframe*, and it is inefficient because approximately 20% of the ADSL symbols are "transitional"; that is, they incur some FEXT and some NEXT, and because of the binary (heavy/light) loading scheme, they must be classified as light.[18] The reader is referred to G.992 for all the details.[19]

How It Should Have Been (Could Still Be?). The sync symbol was devised in 1992 and included in the DMT standard *just in case it was needed* to deal with loss of symbol synchronization. The original cyclic prefix of 40 samples was shortened to 32 [see equation (8.5*b*)], and the saved samples accumulated for 68 symbols to form the sync symbol. Subsequently, as synchronization methods were developed, it was realized that the sync symbol was not really needed. At different times during the discussions on T1.413 Issue 2 there were proposals to remove it, but it was decided that the 1/69 gain in capacity was not worth the inconvenience of backward incompatibility.

When the problem of synchronizing with TDD ISDN's 400 Hz arose, however, there was no need for a strict adherence to T1.413 with its cyclic prefix of 32 samples.[20] The simplest solution would have been to discard the sync symbol and revert to a cyclic prefix of 40 samples[21] (see Table 8.1) with the data symbol rate = the on-line symbol rate = 4 kBaud; there would then be exactly 10 ADSL symbols in one TDD frame. The crosstalk/loading status would repeat every 10 symbols, and there would be no need for any transitional symbols.

A possible timing diagram is shown in Figure 9.12, where, as in G.961, time is measured in unit intervals (UIs) of 3.125 µs, and the propagation on the longest loop (9 km) is assumed to be 50 µs (16 UIs). It can be seen that if the ATU-C transmitter begins its superframe 16 UIs before the ISDN transmitter, the downstream superframe can comprise five heavily loaded and five lightly loaded symbols, and the upstream superframe, four and six, respectively.

---

[17] Probably an influential company had a 32-sample cyclic prefix cast in silicon and was too shortsighted to abandon it: an egregious example of commercial muscle defeating technical merit!

[18] According to Annex C the system is even less efficient than that; out of 345 symbols only 124 are allotted for heavy loading (when there is only FEXT) and 216 for light (when there is any NEXT). I do not understand this.

[19] You can tell my heart is not in it!

[20] There would be no problems of end-to-end compatibility here; if one ATU has to coexist with TDD ISDN, so does the other.

[21] With a small reduction in distortion as a spin-off benefit.

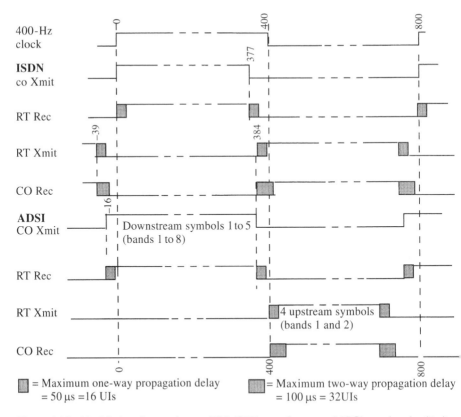

Figure 9.12   Ideal timing diagram for one TDD ISDN superframe, and ADSL synchronized to it.

9.3.2   Band Assignments and FFT Sizes

Because the crosstalk is the same for downstream and upstream, the channel capacity is also. The original proposal was to use the full bandwidth for both directions in a TDD mode, but because (1) the upstream rate need be only one-eighth of the downstream, and (2) Annex A of G.992 defines a conventional FDD system with band 1 allotted to upstream and bands 2 through 8 allotted to downstream, it was decided to stay with this band assignment and use dual bit loadings in both directions.

How It Might Have Been. Annex A FDD with dual bit loading is inefficient because

- During ISDN up the downstream capacity in band 2 (light loading) ≪ the upstream capacity (potential heavy loading)
- Conversely, during ISDN down the upstream capacity in band 1 (light loading) ≪ the downstream capacity (potential heavy loading).

TABLE 9.4   Recommended Band Assignments for ADSL with TDD ISDN Crosstalk

During	Mandatory	Optional	Perhaps
ISDN up	Upstream in band 1	Upstream in band 2 (note 1)	Downstream in bands 3–8 (light loading) (note 2)
ISDN down	Downstream in bands 1–8		

Notes: 1. Annex A requires only a 64-pt (I)FFT, and Annex B *could* be implemented that way, but for all the reasons discussed in Section 8.3.4, a 128-pt (I)FFT will probably become the norm.
2. Some existing upstream ISDN receivers may have very little receive filtering and therefore not be able to tolerate the high levels of out-of-band NEXT caused by ADSL downstream transmission against the flow.

Therefore, considerably higher data rates could be achieved by the assignments shown in Table 9.4. Thus dual bit loading would never be needed for upstream. For downstream it should be an option[22] (controllable by system management); its advantages (A) and disadvantages (D) are as follows:

D. If quad separation is not implemented, the small increase in capacity achieved by transmission against the flow in bands 3 through 8 may not be worth the additions to the training sequence that would be needed to calculate and transmit the extra bit table.

A. If the TC and higher layers are able to take advantage of periods of higher data rates, dual bit loading should be retained for those times when there is very little or no active ISDN in the cable: the light becomes much heavier.

D. Without dual bit loading in either direction the system becomes pure TDD, which could be implemented cheaply with a single (I)FFT processor in each ATU.

9.3.3   Separate Quads for ISDN and ADSL

If an overall DSL system could be managed so as to put ISDN and ADSL in separate quads, severe same-quad alien crosstalk could be avoided, and the worst-case alien XT coefficents would be lower than for unquadded cable; dual bit loading for downstream would be more advantageous.

9.3.4   ULFEXT from Close-in ISDN Modems

Usually, the upstream ADSL capacity will be greater than one-eighth of the downstream because the upstream uses two bands instead of the usual one. If, however, some of the ISDN loops are short, there is a potential for high levels of

---

[22] The training sequence should be arranged so that if one of the ATUs does not have the capability, the other should use a zero second loading.

FEXT into upstream ADSL because TDD ISDN does not use any power cut back. The danger of this effect would seem to be a strong argument for separate quads, but separation may occur naturally without any conscious management: loops in the same quad would probably be of similar lengths.

10

VDSL: REQUIREMENTS AND IMPLEMENTATION

VDSL is not as mature as ADSL, so in this chapter I can only define the system requirements, and describe and compare *proposals* for the duplexing and modulation. The ITU will probably choose one (or a combination) sometime after I submit the manuscript of this book (March, 1999) and before its publication.[1] The process by which the choice will be made is very controversial. For ADSL the choice was based on the results of competitive tests, but sentiment is running against these for VDSL. The ideal alternative would be complete disclosure by all proponents of both the theory and practice of their method followed by objective analysis and discussion. I know that to hope for this would be naively unrealistic; corporate alliances, investment in committed silicon, intellectual hubris, managerial incompetence, backroom cajoling and arm-twisting, shortsightedness, and a difference of interest between the telcos and the modem manufacturers will all play a part.

There are three candidates for the duplexing method: FDD, an interesting technique called Zipper, which is a way of doing FDD with a minimum of filtering, and S(ynchronized)TDD; these are discussed in Sections 10.3 through 10.5. It is unlikely that STDD will be chosen because of the telcos' fear that it will be difficult (perhaps occasionally impossible) to synchronize all the VTUs in a cable.[2] Nevertheless, some form of synchronized DMT will be needed for a VDSL system that has to be binder-group compatible with TDD ISDN (see Section 9.3); a form of Zipper has been proposed for this also. In the last section of this chapter I will try to be unbiased[3] and provide an objective comparison between the two pairs of candidates: FDD and Zipper for "general-purpose" VDSL, and SDMT and Zipper for ISDN-synchronized VDSL.

There are also three candidates for the modulation method: single-carrier QAM and CAP, and DMT, but I think readers will empathize if I discuss only the last one. All three are discussed in [Cioffi et al., 1999]: a good survey paper. This

---

[1] The worst possible timing: the reader will know more than the author!
[2] More technical discussion of this in Sections 10.5.3 and 10.7.12, and a personal opinion in Section 10.8.
[3] I originally proposed it for VDSL, and I still believe it would be the best, so that will be hard!

chapter is much less detailed than Chapter 8 on ADSL for two reasons: many of the implementation details are similar to ADSL and need not be repeated, and many other details have not yet been worked out.

10.1 SYSTEM REQUIREMENTS AND CONSEQUENCES THEREOF

The requirements are completely defined[4] in [Cioffi, 1998] and [ETSI, 1998]. For our purposes the six most important are (1) the services/ranges/rates combinations; (2) transmit PSDs; (3) compatibility with wireless systems, particularly AM and amateur ("ham") radio; (4) coexistence with ADSL; (5) operation on the same pair with BRI; and (6) position of the network termination (NT); these are discussed in the following subsections. Some of the consequences of these requirements are common to any method of implementation; these are discussed in this section. Some are specific to either SDMT or Zipper and are discussed in Section 10.4 or 10.5.

TABLE 10.1 VDSL Services, Ranges, and Rates

Service Mode and Loop	Range (kft)	Downstream Rate (Mbit/s)	Upstream Rate (Mbit/s)
Asymmetric			
Short	1 (0.3 km)	52	6.4
Medium	3 (1 km) (note 1)	26	3.2
Long	4.5 (1.5 km) (note 1)	13	1.6
Very long	6 (note 2)	6.5 (note 2)	1.6 or 0.8 (note 3)
"Extended"	(note 4)		
Symmetric			
Very short	note 5	34	34
Short	1	26	26
Medium	3	13	13
Long	4.5	6.5	6.5
Very long	note 5	4.3	4.3
Extra long	note 5	2.3	2.3

Notes:
1. These conversions between kft and km are about 9% inaccurate; it is not clear whether the kft or the km should be paramount. We consider the (shorter) kft ranges in this chapter.
2. The very long 6.5/0.8-Mbit/s service thus overlaps the ADSL service (but with a much shorter range!) I would not dare to predict what this portends for the future coexistence of ADSL and VDSL modems!
3. Intermediate asymmetries of down/up = 4:1 and 2:1 have not been specified, but may be useful later.
4. Interest has been expressed in some countries for a service on a 0.9-mm (approximately 19-AWG) loop as long as 3.4 km. The ITU may have to take this into account in their deliberations.
5. These ranges have not yet been defined.

---

[4] Two important early inputs to the process of developing systems requirements were a conference of mostly European and Asian telcos [FSAN, 1996] and [Foster et al., 1997], in which the "al." was many of the participants in that conference.

10.1.1 Services, Ranges, and Rates

Both asymmetric (with a down/up ratio of $8:1$) and symmetric services should be supported: preferably by a simple reconfiguration of one versatile modem. The various combinations of services, ranges, and rates are defined in Table 10.1. How many of these combinations will have to coexist in the same binder group will depend on the telco/LEC. ADSL loop lengths within one distribution binder group typically do not vary by more than about $\pm 30\%$ about some average, and offering every customer in a binder group the same rate is not very suboptimal. The variance of the shorter VDSL loops, however, will probably be much greater, and service providers will have to decide whether to "preserve the convoy"[5] or to allow different speeds (with all the dangers of collision). One particular symm/asymm combination that was predicted in [Foster et al., 1997] is short-range symmetric (for business customers) and long-range asymmetric (for domestic customers).

The services in a binder group can be summarized:

(a) Homogeneous binder group (all asymmetric or all symmetric)
(b) Mixed binder group (long asymmetric and short symmetric)
(c) Mixed with similar ranges (and therefore aggregate rates) for asymmetric and symmetric

and the expectation is that prob(a) > prob(b) > prob(c). It is desirable (though not essential) that the range/rate for any particular service should not be reduced by the need to provide for a service with a lower probability (i.e., homogeneous gets top priority, etc.).

10.1.2 Transmit PSDs and Bit Loading

The first proposal was for a maximum PSD of $-60\,\mathrm{dBm/Hz}$—modified for compatibility with ham radio (see Section 10.1.5)—across the entire used band. The lower limit of this band was set at 0.3 MHz (see Section 10.1.4), and the upper limit of the usable band for very short loops is about 15 MHz. At $-60\,\mathrm{dBm/Hz}$ the total power in this band (excluding the ham bands) would be about 11.0 dBm; the maximum total transmit power is specified as 11.5 dBm.

PSD "boosts" to as high as $-50\,\mathrm{dBm/Hz}$ outside the ADSL band will probably be allowed, but the total power would still be limited to 11.5 dBm. A mask as shown in [Cioffi, 1998] would allow up to 17.5 dBm if the maximum were transmitted at every frequency, or looked at another way, would limit the bandwidth used to about 1.4 MHz if the maximum allowed PSD were used. Clearly, these two inconsistent specifications will allow a lot of flexibility in the choice of PSD and bit loading.

---

[5] "The speed of a convoy is that of the slowest ship."

If the PSD is limited to $-60\,\mathrm{dBm/Hz}$, the bit loading procedure is very simple: the usable frequency bands and/or time slots for any session are defined by the higher-level system management, and the procedure (for both down and up) to meet a request for a pair of data rates is essentially the same as used for ADSL, which is described in Section 8.5.2. If, however, power boosts are allowed and the dominant constraint becomes the 11.5-dBm total power, then the optimum procedure will be complicated. Over a significant part of the band the receiver noise is dominated by kindred FEXT, so if all VTUs in the binder group obeyed the same rule, the PSD in that part of the band could be reduced below $-60\,\mathrm{dBm/Hz}$ and the savings in total power devoted to the higher frequencies where the noise is AWGN-limited. That is, however, a big "if"; if the final standard allows power boosts to much higher than $-60\,\mathrm{dBm/Hz}$ at any frequency, a unilateral reduction of PSD (i.e., without any assurance that the other VTUs will do the same) in the FEXT-dominated parts of the band would be foolish! If a PSD strategy is not coordinated at a higher level, however, then in the PMD layer each VTU must consider the measured noise as unalterable, and optimize its own transmission under the mixed—total power and PSD—constraint using an algorithm like that in Section 5.3.2.

PSD Cutbacks to Reduce ULFEXT. In Section 4.6.3 we saw if there is a (CO) mix of short and long loops in a binder-group, UL FEXT from a close-in upstream transmitter would be the dominant crosstalk into the upstream on the long loop. This would be particularly troublesome for VDSL if a LEC offered both long asymmetric and short symmetric: clearly the upstream PSD must be reduced.

T1.413 defines a procedure for physical-layer-control of the downstream ADSL PSD, but this mix of VDSL services is much more complicated; it requires system management to decide on an acceptable trade-off between rates, and then control (via a downstream message from the VTU-C) the upstream PSD. At the time of going to press this is *unfinished business*.

10.1.3 Coexistence with ADSL

VDSL and ADSL systems may use pairs in the same binder group in the two ways shown in Figure 10.1: FTTE may result in an *exchange mix* and FTTC may result in a *remote mix*. The potential for crosstalk is different in the two cases.

Exchange Mix. With an exchange mix the ADSL downstream signal may NEXT into the VDSL upstream and/or FEXT into the VDSL downstream. The expected levels for a 4.5-kft VDSL loop are shown in Figure 10.2 over somewhat more than the ADSL frequency band. It can be seen that the (VDSL) signal to (ADSL) NEXT ratio varies between $+12\,\mathrm{dB}$ at $0.3\,\mathrm{MHz}$ and approximately $-11\,\mathrm{dB}$ at 1 MHz; clearly, the entire ADSL band would be useless for upstream VDSL. For the downstream signal the VDSL signal to ADSL FEXT ratio varies between 30 and 18 dB. This is not negligible, but as can be seen from

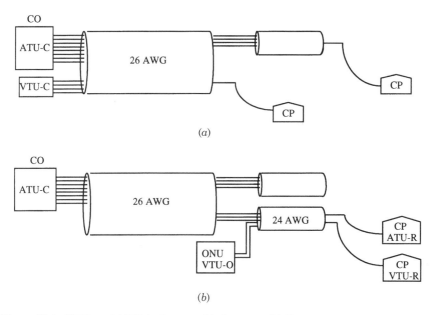

(a)

(b)

Figure 10.1   VDSL and ADSL in the same binder group: (a) fiber to the exchange: exchange mix; (b) fiber to the cabinet: remote mix.

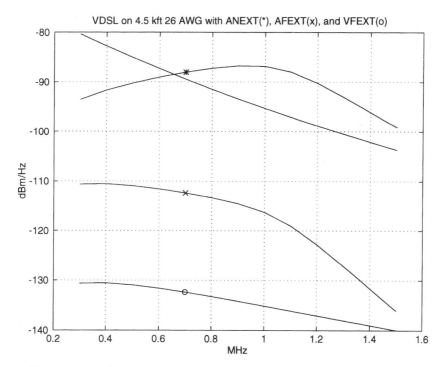

Figure 10.2   VDSL signal and ADSL NEXT and FEXT for 4.5 kft with exchange mix.

TABLE 10.2   Downstream Capacities

	VDSL Downstream Capacity (Mbits/s)	
Loop Length (kft)	In ADSL Band	In Rest of Band
1.0	3.9	58.1
3.0	2.6	22.9
4.5	2.3	7.1

Table 10.2, the contribution to the total downstream capacity varies between small (6.3% of total) and moderate (22%).

Remote Mix. With a remote mix, even though VDSL's PSD is 20 dB lower than ADSL's, NEXT from the upstream and/or ULFEXT from the downstream may be the dominant crosstalkers. Figure 10.3 shows the receive levels of downstream ADSL on a 9-kft 26-AWG loop compared to the two standardized crosstalks, ADSL FEXT and HDSL NEXT, and the two new crosstalks, NEXT from upstream VDSL and UL FEXT from 3 kft of down stream VDSL. It can be seen that beyond about 0.3 MHz, *both directions* of VDSL intefere very badly with ADSL. The downstream ADSL data rates achievable are shown in Table 10.3.

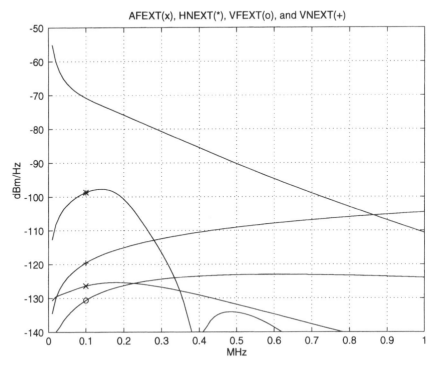

Figure 10.3   ADSL received signal on 9 kft of 26 AWG, with standardized ADSL FEXT and HDSL NEXT, and new VDSL NEXT and UL FEXT from a 3-kft remotely mixed loop.

TABLE 10.3 Downstream ADSL on 9 kft of 26 AWG with XT from Remote Mix VDSL

	D-s ADSL (Mbits/s)	
Interferers	1-kft VDSL Loop	3-kft VDSL Loop
HDSL + ADSL	6.9 (note)	6.9
HDSL + ADSL + VFEXT	3.1	4.9
HDSL + ADSL + VNEXT	2.3	2.3

Note: This is approximately the rate specified in T1.413.

TABLE 10.4 Crude Quantitative Summary of XT Between ADSL and VDSL

Mix	VDSL Down into ADSL Up (see note)	VDSL Up into ADSL Down (see note)	ADSL Down into VDSL Down	ADSL Down into VDSL Up
Exchange	—	—	FEXT	NEXT
Remote	FEXT	NEXT	FEXT	—

Note: These are XTs from a new service into an existing service; they are therefore completely unacceptable.

Summary. The preceding results are summarized in Table 10.4, where the size of the XTer indicates the seriousness of the interference.

Spectrum Management. It is clear from Tables 10.2 and 10.4 that if there is an exchange mix of ADSL and VDSL, then VDSL up *should not* use the ADSL band (0.3 to 1.1 MHz); VDSL down may use that band with a small to moderate benefit depending on the length of the VDSL loop. From Tables 10.3 and 10.4 it can be seen that if there is a remote mix, VDSL *must not* use the ADSL band for either direction. The extent of the control required from the LEC depends on the duplexing method used.

NOTE: Early versions of the requirements document do not recognize any difference between exchange and remote mixes; the presence of any ADSL in any binder group puts the ADSL band off-limits. This seems unnecessarily restrictive; it may be relaxed.

10.1.4 Coexistence with Echo-Canceled BRI

It was agreed from the early days of discussion of ADSL that there would be (at least) two versions: one to operate on the same pair with POTS (to work "over" POTS in the frequency domain), and the other to work over BRI[6]; the extra 140 or so kiloHertz of bandwidth gained if the loop did not have to carry ISDN made the "POTS" version worthwhile.

---

[6] These are defined in Annexes A and B of G.992.

For VDSL, however, this extra bandwidth is insignificant compared to the total bandwidth (about 12 MHz on a short loop), so it was decided from the start that there need be only one version of VDSL[7] with a usable band starting at about 300 kHz. This had the advantage of simplifying all interference calculations because above 300 kHz HDSL NEXT falls off very rapidly, and the only significant alien interferer is ADSL.

10.1.5   Compatibility with Amateur (Ham) and AM Radio

This requires the ability to deal with interference from nearby transmitters (ingress), and in the case of ham radio only, also requires the prevention of interference with nearby receivers (egress).[8] Both ingress and egress are more serious at the remote end, where the coupling is via the (relatively) unbalanced drop wire; an unbalanced and unshielded pair out of a CO or an ONU (a "head end") is less common. Nevertheless, some operators use aerial cable all the way from a head end to the customer premises, so ingress and egress at both ends must be considered.

Ingress. The levels are defined in Section 3.7; methods of dealing with the ingress are described in Sections 10.4 and 10.5.

Egress. The four factors controlling VDSL interference with ham receivers are the VDSL (differential mode) transmit PSD, the differential-mode-to-common-mode balance of the loop near the transmitter, the separation of loop and ham antenna (controlling the path loss), and the sensitivity of the ham receiver. These were all analyzed in [Bingham et al., 1996a], and the conclusion of T1E1.4 was that in the ham bands (defined in Section 3.7) the transmit PSD—both upstream and downstream—should be no higher than $-80\,\mathrm{dBm/Hz}$.

10.1.6   The Network Termination

It is intended that early versions of VDSL service should terminate at a network termination (NT) at the entrance to the customer premises, and that some other method of distribution—as yet unspecified—would be used within the premises. This means that the in-house wiring, particularly the short bridge taps, need not be considered as part of any loop. Whether VDSL will later follow the example of ADSL and be required to operate—at reduced rates if necessary—through existing house wiring remains to be seen.

---

[7] Another difference between the situations for the two systems was that ADSL was developed originally as a standard for North America, where ADSL over ISDN is of little interest. In the development of VDSL, on the other hand, Europe has been the leading partner, and there a mix of ISDN and VDSL on the same pair is important.

[8] Interference by VDSL with AM receivers is not considered to be a problem.

10.2   DUPLEXING

10.2.1   Echo Cancellation?

The first question when considering a duplexing method for VDSL is whether echo cancellation should be used. A strategy for ADSL that is discussed in Section 4.2 is EC up to some changeover frequency and FDD thereafter; for VDSL we can generalize this to consider either FDD or TDD above some frequency. The calculations of Section 4.3 were repeated for the short, medium, and long VDSL ranges (1, 3, and 4.5 kft), but now with only kindred crosstalk as an impairment.[9]

Figure 10.4 shows the aggregate (downstream and upstream) data rates (without any consideration of the methods or efficiency of partitioning) as a function of the crossover frequency. It can be seen that on a short loop, EC up to about 3 MHz would be slightly beneficial (approximately 10% increase over pure xDD), but beyond that the benefits are insignificant. Clearly, EC for VDSL would not be worth the added complexity.

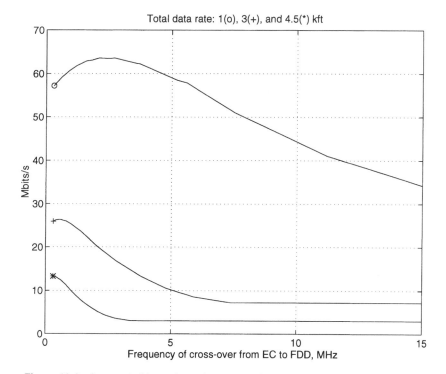

Figure 10.4   Aggregate (down plus up) rates as a function of maximum EC frequency.

[9] The VDSL bandwidth is such that HDSL and ISDN can be ignored. The only potentially serious interferer is ADSL downstream, which we consider shortly.

10.2.2 FDD or TDD?

With EC eliminated, the duplexing choice is between FDD and TDD. Special features and capabilities of FDD, Zipper, and STDD are discussed in Sections 10.3 to 10.5, but motivation for the three proposals, comparison between them, and possible reasons for choosing one or the other are postponed to Section 10.7.

10.2.3 Mixed Services

The basic problem with mixing 8:1 and 1:1 services in the same binder group is that if they are independently optimized, part of the asymmetric's downstream and the symmetric's upstream must share either bandwidth (if using FDD) or time (if using TDD), and will unavoidably NEXT into each other. An extension of the result in Section 10.2.1 is that if—and only if[10]—such a mix is needed, then above some frequency that depends on the loop length both downstream asymm and upstream symm must be constrained so that they do not overlap and do not NEXT into each other. This is fairly easy to do with TDD (see Section 10.6.2) because the capacities of all the symbols in a superframe are approximately equal. It is very difficult to do efficiently with FDD because the capacities of the frequency bands differ widely.

10.3 FDD

The basic principles of DMT FDD have already been discussed under ADSL, and we need only to consider three important differences between VDSL and ADSL:

1. ADSL uses a fixed crossover between downstream and upstream, but it can get away with it because only an 8:1 asymmetry is needed. On long loops the crossover frequency is too high, and the downstream rate is the limiting one, but not seriously so. For VDSL, however, the effect is more severe: for the 1:1 symmetrical service the optimum crossover frequency varies much more with loop length. Flexible crossover frequency(ies) are needed, and modems must be able to change the passbands of all the filters.

2. The requirements for symmetrical operation and for flexibility in choice of downstream and upstream subcarriers mean that both directions must use FFTs spanning the full frequency range; the system considered here uses the 512-pt FFT of ADSL and SDMT.[11]

---

[10] The "only if" follows from the principle established in Section 10.1.1: a homogeneous service must be configured for maximum performance.

[11] For FDD there is no great advantage to be gained from using the very large FFTs required by Zipper.

3. For better spectral compatibility with ADSL, VDSL should not transmit upstream below 1.1 MHz; that is, the downstream band—or at least pat of it—must be below the upstream. Most binder groups will probably not carry both ADSL and VDSL, so this rule would not apply to them, but if the rule can be obeyed without any loss of performance, then for the sake of simplicity, it should be obeyed on all loops.

10.3.1 Mixture of Symmetric and Asymmetric Services

If a mix of $8:1$ and $1:1$ is thought of as a $16:2$ and $9:9$ mix, it can be seen that if all tones were used for both services, 7/16 of the asymmetric downstream tones would suffer kindred[12] NEXT from symmetric upstream; similarly, 7/9 of the symmetric upstream tones would suffer kindred NEXT from asymmetric downstream. This would cause a very severe decrease in the data rates, particularly for the symmetric service.

Just as for SDMT (see Section 10.5.4), for any but the shortest loops, it is a nearly optimum strategy to avoid NEXT altogether by assigning tones exclusively to downstream or upstream.

10.4 ZIPPER

Zipper was described in a series of ANSI and ETSI contributions, [Isaksson et al., 1997], [Bengtsson et al., 1997], [Olsson et al., 1997], and the more accessible [Isaksson et al., 1998b]. The name was derived from the original version of the method, shown in Figure 10.5, in which downstream subcarriers alternate with upstream subcarrriers as in Lewis Walker's invention of 1913.[13] Separation of the transmit and receive signals is achieved by the orthogonality of the DMT subbands.

In FDD the ratio of downstream/upstream data rates is controlled by the frequency bands (both positions and widths) of the filters. Optimizing these filters for a variety of ranges and ratios is a difficult task off line; it is even more difficult on line. Some versions of Zipper control the ratios by using small groups

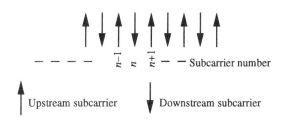

Figure 10.5 Pure Zipper.

---

[12] Cousins rather than siblings.

[13] The patent on that has expired, so there is no problem with IP.

of very narrow subbands in which the number used for downstream and upstream is determined by the downstream/upstream ratio; the prototype system is for symmetrical VDSL, in which alternating tones are used for down and up. This type of Zipper system applied to a channel in which the SNR changes with frequency is an approximate dual of a TDD system applied to a channel in which the SNR changes slowly with time: down/up ratios are controlled by the assignment of successive tones (symbols).

It was recognized, however, that tones near the edges of the bands may suffer from both echo and NEXT (see Section 10.5.3), and changing direction many times across the frequency band increases the number of such tones; more recent versions of Zipper therefore group the down and up tones in large blocks. With this arrangement the name is no longer appropriate, so because it is basically an FDD system implemented without filters the name *digital duplexing* (DD) has been proposed.

The main ways in which Zipper/DD differs from DMT are that it uses:

1. A much longer *cyclic extension* (a cyclic prefix plus a cyclic suffix) to allow for both the receive transient (just as with conventional DMT) and the propagation delay of the loop.
2. A much longer symbol (typically, using a 4096-pt FFT) to maintain efficiency with the longer cyclic extension.
3. Ranging on each UTP in a binder group to ensure that both VTU-O and VTU-R transmitters on any given loop start their symbols at the same time. A method of doing this has not been described, but it is not hard to devise one.[14]

10.4.1 Basic Zipper/DD System

Conventional DMT defines T_{symb} and T_{cp}, the (data) symbol and cyclic prefix durations, and T_{rtran}, the time for the receive transient (after equalization) to decay so that distortion does not add significantly to the noise; the requirement is that $T_{cp} > T_{rtran}$. In addition, Zipper/DD must consider T_p, the one-way propagation delay, T_{etran}, the time for the echo of the transient on another subcarrier on the same UTP to decay so that it does not add significantly to the noise, and T_{Ntran}, the time for the NEXT transient from another UTP to similarly decay. These are shown in Figure 10.6, which shows the timing of transmit and receive signals.

Important points to note are:

- Ranging ensures that transmission starts at the same time at both ends, so the timing diagram (with only the one-way propagation delay accounted for) is the same at both ends.

---

[14] For example, during training the VTU-O could measure $2T_p$ by sending a probing signal, which the VTU-R would immediately return. It would then instruct the VTU-R to start its symbols T_p before the end of the received symbols.

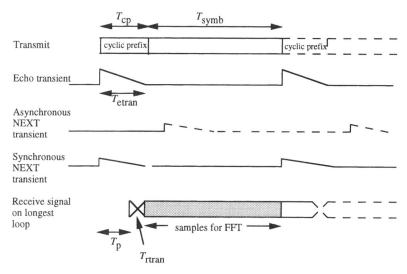

Figure 10.6 Zipper timing diagram on longest loop.

- Echoes need be considered only as "synchronous" transients because transmit and receive use different subcarriers.
- NEXT is also transient, but if symbols on all the UTPs in a binder group are not synchronized,[15] Zipper operates in the *asynchronous* mode, and the transient may occur at any time in the receive symbol. The sidelobes of the full NEXTing signal then impinge on the receive subcarriers, and subcarriers must be assigned in large blocks to minimize edge effects (see Section 10.5.4).
- If all the UTPs in a binder group can be synchronized—when, for example, all the VTU-C/Os are colocated—the NEXT transients occur during the cyclic extension, and their effect is much reduced. Zipper can then operate in a *synchronous* mode, and a more flexible method of subcarrier assignment can be used.

It can be seen that the requirements on the cyclic extension are

$$T_{ce} > T_p + T_{rtran} \qquad \text{for no distortion} \qquad (10.1a)$$

$$T_{ce} > T_{etran} \qquad \text{for no echo} \qquad (10.1b)$$

Equation (10.1a) is just an extension of the $T_{cp} > T_{rtran}$ requirement of conventional DMT, but it adds a little flexibility: if T_{ce} is chosen for a marginally acceptable amount of distortion on the longest loop (the worst case),

---

[15] The most important argument against STDD, which is described in Section 10.6, was the need for synchronizing the superframes on all the UTPs.

distortion will be less on all shorter loops. This will turn out to be very useful when we consider the equalizer in Section 10.4.8. The requirement of equation (10.1b) is considered in the next section.

Efficiency. The formula for the efficiency of Zipper/DD is the same as for DMT: that is,

$$\varepsilon_{\text{Zipper}} = \frac{T_{\text{symb}}}{2(T_{\text{symb}} + T_{\text{ce}})} \qquad (10.2a)$$

$$= \frac{N_{\text{symb}}}{2(N_{\text{symb}} + N_{\text{ce}})} \qquad (10.2b)$$

T_{ce} is determined by the loop (length and IR) and the equalizer, so the efficiency is controllable only by T_{symb}. The values proposed are $N_{\text{symb}} = 4096$ and $N_{\text{ce}} = 320$, so that $\varepsilon_{\text{Zipper}} = 46.4\%$.

Shaping the Cyclic Extension. In Section 6.4 we discussed envelope-shaping of the cyclic prefix. The advantage of shaping is that it reduces the sidelobes, and it can be performed wholly in the transmitter, wholly in the receiver, or partly in both, depending on where the reduction is needed. For Zipper (particularly in the asynchronous mode) the reduction is needed in both in order to reduce the effects of NEXT from nearby subcarriers, and one proposal for shaping was given in [Isaksson et al., 1998a]. The part of the cyclic extension that deals with the propagation delay can be shaped in the transmitter, and the part that deals with distortion can be shaped in the receiver. For the greatest overall reduction of NEXT with a given duration of the cyclic extension, it would be best to have these two parts of equal duration, but with the total cyclic extension proposed in Table 10.5 this is possible only on loops of 4.5 kft or less. The duration of the transmitter shaping can be further doubled by overlapping successive symbols as shown in Figure 10.7.

System Parameters. A preliminary set of parameters is given in Table 10.5. Figure 10.8 shows three plots of the end-to-end attenuation that results from either transmitter or receiver shaping, as a function of the number of tones away from a used band—with no shaping, and with shaping over different numbers of samples: 100 (the maximum for the receiver on the longest loop), 160 (appropriate for both transmitter and receiver on all loops less than 4.5 kft),[16] and 220 (for the receiver on the longest loop). These are the attenuations that would be applied to both echo and ANEXT. Choosing the guard band (i.e., number of unused subcarriers) between a downstream and an upstream band to maximize

---

[16] Isaksson and Mestdagh suggested shaping only 70 of the 220 available in the transmitter, and 70 of the 100 available in the receiver. This reduces the effects of distortion but also reduces the amount of sidelobe suppression. A careful study of the trade-offs is needed.

TABLE 10.5 Preliminary Parameters for Zipper/DD

f_{samp} (MHz)	22.08 MHz
IFFT size (samples)	4096
Overhead for T_p *and* T_{rtran}	320 (14.5 μs)
	(note 1)
Data symbol rate	5.0 kHz
Efficiency	46.4%
Cyclic "suffix" maximum	220
(note 2)	
Cyclic extension (note 3)	$320 + 220$
Subcarrier spacing	5.390625 kHz
Used subcarriers	56–2047
	(0.3–11 MHz)
In-band transmit PSD without	-60 dBm/Hz
power boost	

Notes:

1. An important point about this overhead—not mentioned in [Isaksson et al., 1998a], but clearly premeditated—is that the total of 320 samples is proportionately the same as for ADSL (40/552). This augurs well for "scalable xDSL," an idea that is being developed in SG 15. If the apportioning of samples between transmitter and receiver shaping is changed depending on the length (i.e., propagation delay) of the loop, it is important that this total and the resulting data symbol rate be preserved.

2. This allows for a propagation delay of 10 μs, which is plenty for the longest loop contemplated.

3. The terminology here is confusing, and it would certainly need to be clarified in a standard. The transmit signal comprises $(4096 + 320 + 220)$ phase-continuous samples, of which up to 220 can be used in a shaped prefix and another 220 in a shaped suffix, which is overlapped with the next symbol's prefix: leaving 100 or more to be treated by the receiver as a shapable prefix (independent of the transmitter's prefix). These numbers are incorporated into Figure 10.7, which may help to reduce the confusion slightly.

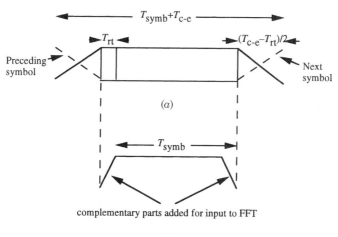

(b)

Figure 10.7 Zipper transmitter (a) and receiver (b) shaping for sidelobe reduction.

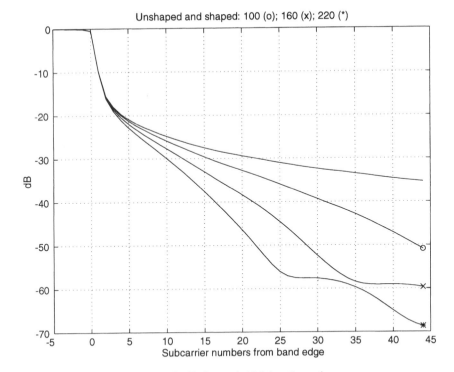

Figure 10.8 End-to-end sidelobe attenuation.

the aggregate capacity would be very complicated, but probably at least 30 subcarriers will be needed. If two downstream and two upstream bands (the minimum for flexibility of services) are used, this would result in an overhead of $90/2048 \approx 4.5\%$; this is small compared to the guard bands of a filtered FDD system, but still not insignificant.

Clearly, the total cyclic extension would have to be defined in a standard, but how much more is a very vexed question. Since the shaping on one UTP affects NEXT into others, it is arguable that it should be defined, but more likely is a set of out-of-band PSD specifications just as for any conventional FDD system. How the transmitter and receiver would then agree on shapings needs further study.

10.4.2 Analog Front End and ADC

Much of the appeal of Zipper/DD lies in its purported ability to perform all the separation of transmit and receive signals with the IFFT and FFT; this brings a lot of flexibility in configuring mixed systems (see Section 10.5.4). This assumes, however, that the full signal at the output of the 4W/2W hybrid (receive signal plus reflected transmit signal) is analog-to-digital converted; this assumption needs to be examined.

With a transmit PSD value of $-60\,\mathrm{dBm/Hz}$ into a 4.5-kft loop of 26 AWG, the used band is approximately 0.3 to 3.0 MHz, and the transmit and receive powers in a symmetric system using "alternating" Zipper[17] are approximately $+2.0$ and $-31.0\,\mathrm{dBm}$. This means that to reduce the echoed signal just down to the level of the receive signal—thereby doubling the power input to the ADC and requiring an extra half of a bit of quantization for the same level of quantizing noise—a *trans-hybrid loss* (THL) of 33 dB is needed.

Without bridge taps and line coupling transformers, and in a frequency region where the range of characteristic impedances of the cables used is small, this would not be difficult. The impedances of 24- and 26-AWG pairs, the most common UTPs in the United States, for example, change slowly and smoothly with frequency, and an RRC matching impedance that is a compromise for the two gauges can achieve a THL of better than 35 dB across the band. Many other countries, however, use a much wider variety of cables than this.

Transformers, however, typically have a $\pm 10\%$ tolerance on their inductance, and the worst-case THL achievable at the low end of the band is less than 26 dB. I have seen no analysis of the contribution of this high-level echo to the total echo power. Bridge taps present the worst problem. Figure 10.9(*a*) shows test loop VDSL4 defined in [Cioffi, 1998]; Figure 10.9(*b*) shows the return losses at both ends relative to the compromise impedance.[18] It is thus the highest RL that can be achieved without some adaptive hybrid, but at the RT it is abysmally low!

It is clear, therefore, that with the worst reflection and no prefiltering, the number of ADC bits would have to be at least four greater than the number of DAC bits chosen[19]: probably at least 14 bits. It should be noted, however, that if large echoes are dealt with merely by increasing the number of ADC bits, all the responsibility for removing echoes and NEXT falls upon the FFT. The levels of asynchronous NEXT may be such that subcarriers must be assigned in large contiguous groups in order to reduce edge effects (see Section 10.5.3).

Another way of looking at this question was explained to me by Nick Sands.[20] The peak voltage capability of an ADC is pretty much defined by the supply voltages. Therefore, increasing the level of the signal input to the receiver by x dB means that the front-end amplifier gain must be reduced by x dB, thus moving the wanted signal x dB closer to the noise floor (typically between -125 and $-135\,\mathrm{dBm/Hz}$ at the present and foreseeable-future state of the art). Without the reflected transmit signal, the input noise would be dominant (as it should be); with $x = 20$ or more the AFE noise floor becomes the limiting factor over a large part of the band.

---

[17] The calculation for a "block" DD system is much more complicated because of the different bands and bandwidths used for the two directions, so the Zipper system is used as a crude average.
[18] Actually, the impedance match at the RT is so bad that a simple 100-Ω reference impedance is almost as good.
[19] Which number would, of course, depend on any PAR reduction technique used.
[20] Private conversation.

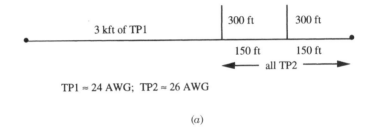

(a)

Figure 10.9 Test loop VDSL4: (a) configuration; (b) return losses at VTU-R relative to Z_0 and 100 Ω.

NOTE: This was one of the reasons for the pre-echo canceler in ADSL, but there it was not so crucial because the input noise was typically much higher than the noise floor.

Reducing the Dynamic Range of the ADC. Fourteen or more bits at a sampling rate of 22 MHz or more may not be feasible, so some reduction of the dynamic range of the ADC may be needed; prefiltering or an adaptive hybrid are possibilities. Prefiltering is possible only if downstream and upstream tones are grouped in large blocks: preferably no more than one block per direction, but with different blocks and different filter passbands for different services and ranges.

Adaptive hybrids have been used with varying degrees of success for a long time. The main problem for xDSL has been achieving the linearity of the digitally programmable analog elements at the xDSL frequencies. [Pécourt et al., 1999] have reported THL values of 25 to 30 dB at ADSL frequencies even with very demanding bridge taps. Whether this can be extended to VDSL frequencies remains to be seen, and much work will be needed to devise line-probing and parameter-calculation methods that can be implemented on line.

10.4.3 Echoes and NEXT

Echoes and NEXT in Zipper are similar to conventional echoes and NEXT in that they involve a high-level transmit signal coupling into an attenuated receive signal: via the 4W/2W network for echoes and via interpair coupling for NEXT. They are unconventional, however, because they come from other subcarriers, and their steady state is zeroed by the orthogonality between subcarriers; consequently, they are only transients. As shown in Figure 10.6, echoes are *synchronous*; that is, transients from the change of the transmit signal (in the same pair) occur during the cyclic prefix (guard period), and most of them will have decayed before the receive samples must be collected. The following preliminary discussion of the magnitudes of these may encourage others to do a more precise analysis.

Echoes. Experience with ADSL has shown that *if* the precanceler reduces the echo enough that the SER at the input to the $ADC > 0\,dB$, echoes out to about 150 μs may contribute significantly to the total noise. VDSL systems scale all ADSL frequencies and times by a factor of approximately 10, so if comparable echo levels are achieved either by a good THL (e.g., on loops without bridge taps) or by prefiltering, we might expect that echo levels would be significant out to 15 μs. Zipper echoes, however, are on other subcarriers—either adjacent or far removed, depending on how the subcarriers are assigned—and will be reduced by the sidelobe attenuation; allowing for $T_{\text{etran}} \approx 10\,\mu s$ would probably be adequate. As we shall see in Section 10.7, in these cases requirement (10.1a) appears to be dominant. If, however the THL is low because of bridge taps (e.g., Figure 10.9) and prefiltering is not used, careful calculations will be needed for assurance that T_{etran} is indeed less than T_{ce}.

Synchronous NEXT (SNEXT). This is similar to echo in that it is a transient that generates only interchannel "noise," but it will usually be much lower in level than the echo. SNEXT, however, is not attenuated by the 4W/2W network, and furthermore, the plots of pair-to-pair NEXT in Figures 4 and 5 of [Huang and Werner, 1997] suggest that the IR of a NEXT path may be considerably longer than that of an echo.

SNEXT is *probably* less serious than echo, but more study is needed to prove this.

NOTE: To test this it will be essential that real pair-to-pair NEXT be used. Simulated NEXT with a smooth $f^{1.5}$ spectrum will have an IR that is unrealistically short.

Asynchronous NEXT (ANEXT). [Isaksson et al., 1998a] showed simulation results with (ANEXT) when downstream and upstream are assigned alternating blocks of 200 subcarriers (2048 total). As would be expected, ANEXT is the dominant impairment (i.e., greater than kindred FEXT) at the edges of each of the bands, and the reduction of capacity will be small to moderate, depending on the number of bands used (nine in their simulation, but ideally four or fewer).

10.4.4 Mixture of Symmetric and Asymmetric Services

Just as for conventional FDD considered in Section 10.3.1, Zipper must provide flexibility in the assignment of passbands. The tactics for doing this may be very different for synchronous and asynchronous Zipper.

Synchronous Zipper. This provides almost as much flexibility as SDMT (see Section 10.5.4). A first try at a compromise assignment of tones suggests that out of every $18n$ tones the asymmetric service should use $11n$ down, $2n$ up, and $5n$ quiet, resulting in a downstream data rate that is approximately 0.69 of that achievable with homogeneous binder groups; the symmetric service should use $7n$ down, $7n$ up, and $4n$ quiet, for a downstream/upstream rate that is approximately 0.77 of the "homogeneous" rate.

NOTES:

1. n does not have to be equal to 1. For ease of bit loading it is necessary only that the SNR vary no more than 3 dB across each block of $18n$ tones.
2. This tone assignment should be done by spectral management from a higher layer based on what services are *available* in the binder group (not what are being used at the time, because that might require too much DRA when those services do come on); the decision cannot be left to the PMD layer because that would result in a "first up takes all" result.

Asynchronous Zipper. If the symbols within the binder group are not synchronized then the problem is the same as with a conventional FDD system: the downstream and upstream tones should be in as few contiguous groups as possible in order to minimize edge effects.

10.4.5 Coexistence with ADSL

As suggested in Section 10.3, unless it can be shown that there is some big advantage to be gained otherwise, the block of tones up to 1.1 MHz should be reserved—by system management regardless of whether there is ADSL in the binder group or not—for downstream.

10.4.6 Coexistence with TDD BRI

Although a BRI signal is only mildly bandlimited, it does eventually roll off, and above some frequency (3 to 6 MHz, depending on the length of the loop) NEXT from it becomes smaller than VDSL's kindred FEXT. One variant of Zipper/DD takes advantage of this to reduce the latency slightly by dividing the usable band into two parts. In the lower part, where alien NEXT would be significant, it transmits "with the flow." In the higher part, where only kindred NEXT would be significant *all* VTUs transmit against the flow. It is thus a synchronized DMT system, and might be characterized as a hybrid TDD/FDD. It is slightly inefficient because of the extra band splitting needed and because alien NEXT is never zero.

10.4.7 Bit Loading

If the rate ratios and the bands to be used are defined by a higher layer, and the maximum PSD is also fixed, the bit loading procedure can be similar to that defined for ADSL in Section 8.5.2. If, however, power boosts are allowed, and the PSD limits become incompatible with a total power limit, then, as discussed in Section 10.1.2, the procedure will be very complicated.

10.4.8 Equalization

Short loops have higher SNRs on most of their subcarriers, so they require higher SDRs. They also have shorter propagation delays, so the part of the cyclic extension that was allotted to deal with the delay can be shortened, and the part that deals with distortion can be lengthened, thus taking some pressure off the equalizer; a much simpler TEQ should be acceptable.

NOTE: This trade-off between T_p and T_{rtran} can be done during initialization after the ranging. T_{cp} can be adjusted and all subsequent equalizer training and SNR measurements done with the longer T_{cp}.

10.5 SYNCHRONIZED DMT

A synchronized TDD system was proposed to ANSI and ETSI [Bingham, 1996] and [Bingham et al., 1996b]. The basic principle—alternating bursts of downstream and upstream transmission that are separated by short quiet periods to allow for the propagation delay of the loop—has been used for many years in STDD ISDN as defined in G.961 Annex 3.

One of the big merits of a synchronized (SDMT) system is that it is pure TDD, and each transceiver needs only *one* (I)FFT core, which, with trivial changes

TABLE 10.6 Basic Numbers for SDMT VDSL

f_{samp}	22.08 MHz
IFFT size	512
Cyclic prefix	40
Data symbol rate	40 kHz
Subcarrier spacing	43.125 kHz
Used subcarriers	7–255
	(0.3–11 MHz)
In-band transmit PSD without power boost	−60 dBm/Hz

(see Appendix C), serves alternately as IFFT and FFT: a *big* saving in power and either silicon or DSP cost. The proposed sampling and symbol frequencies, which are shown in Table 10.6, are exactly 10 times those used for downstream ADSL[21]; the symbol duration is thus 25 µs.

We shall see in Section 10.7 that this system is superior to FDD and Zipper in nearly all respects. The important exception—the Achilles heel of SDMT[22]—is the need for synchronization of all the superframes in a binder group.[23] It is therefore unlikely that SDMT will be standardized for most VDSL systems. If, however, a binder group already carries TDD ISDN signals, the NEXT from these into an unsynchronized VDSL system would exceed the levels of kindred FEXT at frequencies below about 3 MHz and would therefore seriously reduce the VDSL capacity. The solution, clearly, is to synchronize a TDD VDSL system to the 400-Hz clock that is already being used for the ISDN.

NOTE: Even with synchronization, because TDD BRI is a symmetrical system, only symmetrical VDSL will be completely free from aNEXT.

10.5.1 Basic DMT System Compatible with TDD BRI

In addition to T_{symp}, T_{cp}, and T_p, which have been already defined, SDMT requires:

- T_{sf}, the duration (down plus quiet plus up plus quiet) of a superframe; this must be equal to the ISDN superframe: that is, 2.5 ms or $100 \times T_{symb}$.
- T_q, the total quiet time per superframe: must be $>2T_p \approx 20$ µs.

Each quiet period should, for simplicity of implementation, be an integer number of DMT symbols ($= 25$ µs), so the minimum value of T_q is 50 µs (much more than is actually needed), leaving 98 symbols for data. The superframe of a symmetrical system should therefore be configured as 49d/1q/49u/1q.

---

[21] The data rates of asymmetric VDSL (56/6.4) are also approximately 10 times those of ADSL.
[22] The recommendation in [FSAN, 1999] was the fatal arrow.
[23] Synchronous Zipper also needs this, but synchronization is only icing on the cake for Zipper; it is bread and butter for SDMT.

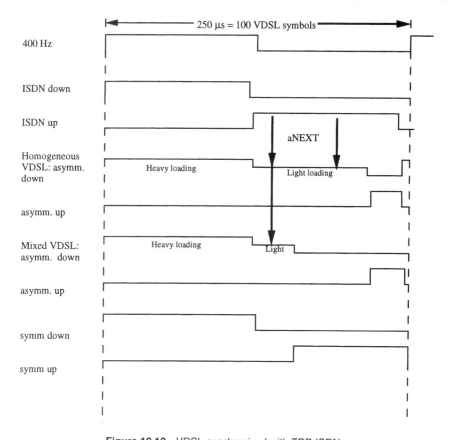

Figure 10.10   VDSL synchronized with TDD ISDN.

A homogeneous asymmetrical system needs more careful calculation because slightly less than half of the downstream symbols would incur alien NEXT (aNEXT), and the optimum superframe format would depend on the loop length; my first estimate for the superframe format would be 90d/1q/8u/1q, as shown in Figure 10.10. It can be seen that the downstream must use dual bit loading [Chow and Bingham, 1998]: heavy during the 50 kFEXT symbols, light during the 40 (kFEXT + aNEXT).

A heterogeneous (mixed symmetrical and asymmetrical services) VDSL system requires yet another set of superframe formats because there is now the potential for kindred NEXT. We have previously seen that above about 3 MHz, kNEXT makes duplex transmission (simultaneous downstream and upstream) impractical; now, below about 3 MHz, aNEXT from TDD ISDN makes it impractical. Therefore, no duplex transmission should be used. Figure 10.10 shows the superframe formats without quantifying the segments; more work is needed.

Efficiency. The overall efficiency of SDMT is the product of the basic DMT efficiency and the duplexing efficiency of equation (4.7):

$$\varepsilon_{\text{SDMT}} = \frac{T_{\text{symb}}}{T_{\text{sumb}} + T_{\text{cp}}} \times \frac{1 - T_{\text{q}}/T_{\text{sf}}}{2} \qquad (10.3a)$$

$$= \frac{512}{552} \times \frac{98}{100}$$

$$= 45.4\% \qquad (10.3b)$$

NOTE: This number is about half those cited in previous descriptions of SDMT, but it is consistent with the convention established in Chapter 4 that only a perfectly echo-canceled system is 100% efficient.

Unfinished Business. The rules (or guidelines, really) for spectral management that are proposed above are both complicated and immature; more work may be needed before the full "capacity" of a heterogeneous VDSL system coexisting with TDD can be achieved.

10.5.2   Analog Front End and ADC

Because SDMT uses TDD with a quiet period between transmission and reception, the ADC can be turned off during transmission, and the 4W/2W network is unimportant. The front-end problems are *much* simpler than for FDD systems.

10.5.3   Synchronization

The difficulties of synchronizing all SDMT systems in a binder group notwithstanding, there is no alternative for VDSL coexisting with TDD ISDN. The task has three parts:

1. All VTU-C/Os must synchronize their superframes to the ISDN TDD in order to prevent both alien and kindred NEXT.
2. All VTU-Rs must lock their 22.08-MHz sampling clock to the received downstream clock, and then synchronize their 2-kHz superframe clock (and thereby also their 40-kHz symbol clock).
3. Each VTU-O or C must then delay its receive superframe and symbol clocks to match the upstream signal.

We must consider all possible solutions to these problems, but if some of them seem simplistic, it is because they are envisioned by a modem designer; they will need refining by a *motivated* systems designer.

Synchronizing Co-located VTU-C/Os. Racks of VTU-Cs or VTU-Os will usually include a master oscillator. This should pass both a 22.08-MHz sampling clock and a 400-Hz superframe clock to each VTU-C or O. If it is not possible to synchronize the 400-Hz clocks delivered to different racks, these should be treated as separated VTU-C/Os (q.v.)

Locking and Synchronizing VTU-Rs. Recovery of the sampling clock is done using a pilot just as with ADSL (see Section 8.4.3). Recovery of the superframe and symbol clocks is easier than with ADSL because of the burst format. The following description is from [Sands and Bingham, 1998]. The preferred place for any modem timing recovery is at the front end of the receiver (feedforward recovery), and indeed, a DMT time-domain signal does contain some timing information because the samples in each cyclic prefix are correlated with those at the end of each symbol. High-level narrowband RFI, however, obscures this information, and timing recovery can be performed only after the RFI has been filtered out: that is, at the output of the FFT (feedback recovery).

A simple and robust measure of the alignment of any superframe and symbol timing is the sum of the powers on all subcarriers that have a high SNR (i.e., leaving out all those that are contaminated by RFI or any other high-level noise). Figure 10.11(a) shows a received symmetric (9/1/9/1) superframe, an unconverged receiver timing that is initially 1.75 symbols (1 symbol and approximately 400 samples) late, the sum of the powers [$p(k)$ for $k = 1$ to 20] on all the "clean" subcarriers and their first finite differences, $\Delta p(k)$ $= p(k) - p(k - 1)$. Figure 10.11(b) shows the same for a timing that is initially 2.25 symbols late. It can be seen that the first symbol should be that with the maximum positive value of $\Delta p(k)$, and that the optimum sample timing results in $\Delta p(k - 1) = \Delta p(k + 1) = 0$. The symbol index should therefore be reset (in both cases advanced by two), and then the timing should be delayed by one sample if $\Delta p(2) > \Delta p(20)$, or advanced if $\Delta p(20) > \Delta p(2)$.

Because the FFT uses only 512 of the 552 receive samples, this algorithm has a "flat spot" when the beginning of the sampling is within the cyclic prefix; the sum of the output powers would be approximately the same for about 39 sampling phases. This can be remedied by collecting the 512 samples alternately with $k = 1$ to 512 and $k = 41$ to 552 and then adding the two successively calculated power sum differences. Thus, if one of the first samples is within the cyclic prefix, the other will not be, and [$\Delta p(k) + \Delta p(k + 40)$] will change— albeit at half the rate—with each sampling shift.

Delaying the VTU-C/O Receiver Clock. The upstream sample timing can be retrieved by a simplified version of the previous algorithm: because the round-trip delay is less than one symbol period, the transmit and receive symbol indices are the same, and no initial resetting is needed.

Synchronizing Separated VTU-C/Os. Task 1 above becomes much more difficult in "separated" systems; we must consider two situations:

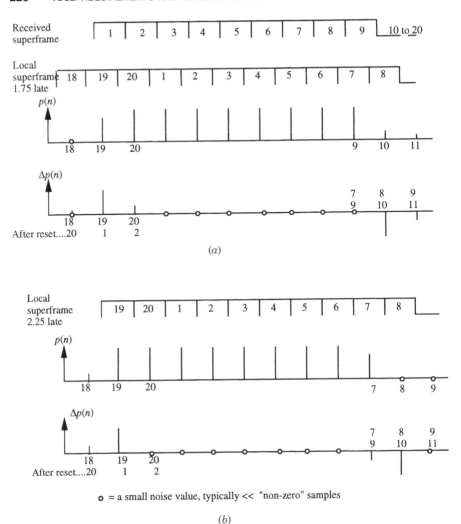

Figure 10.11 Received and unconverged local timing: (*a*) initial timing 1.75 symbols late (0.25 periods early after reset); (*b*) initial timing 2.25 symbols late.

1a. Some "VTU-Cs" are operated by CLECs and are outside the CO.

1b. The VTU-Os are in more than one cabinet.

An external VTU-C does not have access to either of the master clocks: sampling or superframe. It may use its own crystal-controlled sampling clock but must synchronize its 400-Hz superframe clock—with occasional stuffs or deletes of one sampling cycle if needed—with that used in the binder group. Ways in which this could be done include:

1a1. A pilot tone modulated at 400 Hz (either on-off or phase-reversed) could be injected via a high-impedance (current) source at the cross-connect point inside the CO.

1a2. A newly turned-on external VTU-C could listen to NEXT from both ISDN and other, already synchronized VDSLs, and establish its superframe timing using the algorithm described above for the VTU-R. This method would encounter a problem if an external VTU-C were the first up; it would have no established superframe clock to slave to. A solution to this problem has been successfully simulated, but not demonstrated on real loops. More work is needed here.

Ways in which separated VTU-Os could be synchronized include:

1b1. The responsibility for detecting alien (i.e., ISDN) NEXT and locking a superframe to it would have to lie with the VTU-Rs.

1b2. Superframe timing could be included as a marker with the other control information that must be transmitted downstream on the fiber.

1b3. The 8-kHz NTR, which is used throughout the network, could be divided by 4: the problem would thus be reduced from finding a phase for the 2-kHz superframe out of a continuum to finding one of four.

NOTE: None of these methods has been proven, and the last two would require a greater level of cooperation between the PMD and higher layers than heretofore, but the benefits would be great.

Another way that appears to be LEC independent, and might be useful for both situations, is that once installed and first trained, *all* VTU-Os should transmit some or all of the "potential pilot" tones (i.e., numbers 64, 128, 192, and 256, which maintain their phase across the cyclic prefix) on-off keyed at the 2-kHz rate.[24] Any newly installed VTU-C or O would listen for NEXT on one or all of these tones,[25] lock its 2-kHz clock to that, and then start transmitting the "pilots." An ATU-R should not transmit anything until it has locked itself to these.

10.6 DEALING WITH RFI FROM HAM AND AM RADIO

The three ways of suppressing RFI—analog cancellation, filtering, and digital cancellation—are almost equally applicable to SDMT and Zipper, so they are discussed here in a separate section with only occasional references to the

---

[24] This is similar to the "always on" mode that was originally considered essential for G.992.2 but was abandoned because of time pressures. It may be time to revive it.

[25] There is enough diversity that the probability that not one of the "pilots" on any of the already turned on VTU-C/Os will generate detectable NEXT on the new pair is negligible.

duplexing method. As mentioned in Section 3.7, the discussion and publications so far have concentrated on the effects of ham radio RFI on VDSL, so that will be our main concern here. The effects of AM radio will be discussed briefly in Section 10.6.5 (Unfinished Business).

10.6.1 Front-End Analog Cancellation

The worst-case differential-mode RFI signal level at the input to a receiver may be as high as 0 dBm. By contrast, on a long VDSL loop (4.5 kft of 26 AWG) the wideband receive power is approximately -25 dBm. This means that even if perfect RFI cancellation were possible digitally, the 0-dBm analog RFI would have wasted more than four ADC bits. These bits are expensive, so at least 25 (preferably 30) dB of cancellation in the analog domain is needed.

Figure 10.12 shows a generic front-end canceler, as originally described in [Cioffi et al., 1996], in which all operations are performed in the analog domain. The common-mode signal can be very easily detected without loading the receive signal by using a high-impedance center tap on the line side of the transformer. The signal from this should be attenuated by approximately 30 dB in order to bring the gain of W into a reasonable range. Then the adaptation operates to drive the correlation between the "error" signal[26] and the canceling signal to zero.

The ideal situation for such adaptive cancellation of the noise part of a received signal $(y + n)$ is that there be another source that is pure noise; that is, the noise to signal ratio[27] (NSR) $= \infty$. This is nearly achieved here: the signal part of the common-mode signal is typically 20 to 30 dB below the differential-mode signal, so the NSR > 50 dB. Therefore, the step size can be fairly large, and the adaptation fairly rapid.

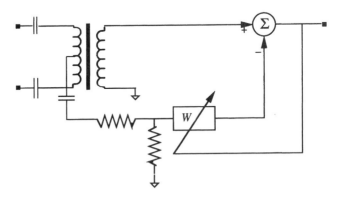

Figure 10.12 Front-end analog RFI canceler.

---

[26] Note that as in all noise cancellation schemes, the "error" that drives the adaptation is the wanted signal.

[27] The unconventional NSR is used here to emphasize that for cancellation the noise is really the "signal" and the signal the "noise"!

The transfer function from ham antenna to xDSL front end changes fairly slowly with frequency, and a single ham band is relatively narrow ($< 4.0\,\text{kHz}$); therefore, the simplest form of W can be just a single complex multiplication. It may be tempting to use more coefficients (taps) for W so as to better fit the variation of the coupling across the band, but this may be counterproductive; if a ham operator changes carrier frequency during a session, a single coefficient may need to be adjusted only slightly, but a close fit across the 4-kHz sideband of one carrier may extrapolate poorly to the sideband of another.

The analog components needed to perform the multiplications involved in adapting and implementing the transfer function W are very difficult to build at VDSL frequencies, so some combination of analog, A/D/A conversion, and digital operations is needed. Early field trials of a hybrid implementation (part analog, part digital) have worked well, and it appears that the 0-dB goal for signal to RFI ratio at the input to the AGC can be attained.

10.6.2 Shaped Windowing

A conventional DMT receiver uses a rectangular "window" to select just N of ($N + \nu$) samples (512 of 552 for SDMT). As we saw in Section 6.4, the sidelobes beyond the first few can be significantly reduced by complementarily weighting the cyclic prefix and the last v samples with a sine-squared (a.k.a. raised-cosine) function. If this is applied only in the receiver (method (2) in Section 6.4) the effects of RFI on tones distant from the radio bands is much reduced. This reduction is not sufficient by itself, but shaping can be combined with either filtering or cancellation.

10.6.3 Digital Filtering

RFI suppression could perhaps be achieved by an adaptive band-reject filter. A simple very narrowband fourth-order IIR section with

$$F(z) = \frac{z^2 + 2\cos\theta_{z1}\,z + 1}{z^2 + 2(1 - \varepsilon)\cos\theta_{p1}\,z + (1 - \varepsilon)^2} \frac{z^2 + 2\cos\theta_{z2}\,z + 1}{z^2 + 2(1 - \varepsilon)\cos\theta_{p2}\,z + (1 - \varepsilon)^2}$$

$$(10.4)$$

where

$$\theta_{z1} = \theta_c - d\theta_z \qquad \theta_{z2} = \theta_c + d\theta_z \qquad \theta_{p1} = \theta_c - d\theta_p \qquad \theta_{p2} = \theta_c + d\theta_p$$

$$(10.5)$$

would probably suffice. The displacement of the zeros from the center frequency, $d\theta_z \approx 8\,\text{kHz}$ for a 4-kHz-wide ham signal; the displacement of the poles, $d\theta_p$, would be a compromise between rejection of the RFI and distortion of the DMT signal. The filter would therefore, have just one adaptable parameter, θ_c, which controls its center frequency.

I have not seen any analysis of the distortion of a DMT signal introduced by such a filter, but judging from the fact that attention seems to have been concentrated on cancellation rather than filtering, the expectation seems to be that the distortion would be unacceptable. It is interesting to speculate on how suppression of RFI would be achieved with SCM.[28] It would appear that adaptive filtering (and its associated distortion and necessary adaptive equalization) is the only method.

10.6.4 Digital Cancellation

DMT-based cancellation of RFI has been described in a series of papers (see [RFI1-5]). It makes use of the fact that tones within the ham bands must be silenced (i.e., "unused") to meet the egress requirements. The steps of the process are as follows:

1. *Off-line.* Model the effects of a narrowband signal centered[29] about f_c on the outputs of the demodulating FFT at all tones on both sides of f_c.

2. *On-line, symbol by symbol.* Calculate f_c and the other parameters of the model by fitting to the FFT outputs on two of the unused tones within the radio band.

3. Calculate the estimated effects of the RFI on the used tones from the now quantified model, and subtract them.[30]

The above steps could be performed using the conventional rectangular window in the receiver, but the slow decay of the sidelobes of the FFT transfer function would have two disadvantages;

- The negative frequency components of the detected interference make it very difficult to model accurately the effects on the FFT outputs.

- *Even if* the effects could be modeled accurately, step 3 above would have to be performed on all the tones in order to achieve the desired 60 dB of cancellation.

It is better to use a shaped window as described under method (2) in Section 6.4. Because shaping increases the effects of channel distortion, it is best to use only as much as is needed to achieve the desired amount of cancellation. For SDMT using a sine-squared shaping over just 20 of the 40 samples of the cyclic

---

[28] For equalization SCM is definitely more mature, but for RFI suppression the reverse is true; MCM is more advanced (not mature, but certainly adolescent).

[29] Note that because the signal is SSB, this is not the "carrier" frequency.

[30] The residue on the unused tones that were not used to calculate the parameters can be used as an indication of the accuracy of the model.

prefix—with, of course, the complementary shaping at the end of the pulse—appears to be a good compromise.

Model of Narrowband Interference. The simplest, zero-order, time-domain model is that of a sine wave with frequency, amplitude, and phase that are constant throughout the symbol period. With SDMT parameters (with a 1:10 ratio of the 25-μs symbol period to the shortest cycle of the 4-kHz RF signal) this model is quite good, and early work indicated that 40 to 50 dB of RFI suppression can be achieved. With Zipper parameters (symbol periods four or eight times longer than with SDMT), however, the approximation is poor.[31]

A first-order model is a sinusoid with constant frequency and phase but with an amplitude that is a linear function of t (i.e., the envelope is trapezoidal). That is,

$$rfi(t) = \text{rect}(t)(a + bt)\cos(2\pi f_i t + \theta) \tag{10.6}$$

where

$$\text{rect}(t) = \begin{cases} 1 & \text{for } 0 < t < T_{\text{symb}} \\ 0 & \text{otherwise} \end{cases} \tag{10.7}$$

This model thus has four parameters: f_i, θ, a, and b.

NOTE: f_i is not the ham carrier frequency, which will typically be constant over many symbols, but rather, a pseudo center frequency—displaced 1 to 2 kHz from the carrier frequency because of the SSB modulation—that will change from symbol to symbol.

The negative frequency components of the response of the "shaped" FFT to this signal can be ignored because the first ham band begins at 1.8 MHz, which is approximately SDMT tone 42, and the magnitudes of sidelobes 84 and greater are negligible; nevertheless, even with this simplification the response is still complicated. One way to make it tractable is to approximate it as the product of the response of the unshaped FFT and a "shaping" factor, W.

The response of the unshaped FFT has two terms: one proportional to $1/(f - f_i)$ that is due primarily to the constant part of the sinusoid, and one proportional to $1/(f - f_i)^2$ that is due to the part that varies linearly with t. If the RFI occurs between tones n and $(n + 1)$, it is convenient to define both f_i and f in multiples of the tone separation:

$$f_i = (n + \delta)\Delta f \tag{10.8a}$$

$$f = (n + m)\Delta f \tag{10.8b}$$

---

[31] Another (the dual?) way of considering this is that cancellation is a process of extrapolation from a measured band to a wider influenced band; with Zipper the measured band is one-fourth or one-eighth of that for SDMT.

so that $(f - f_i)$ can be replaced by $(m - \delta)$. Then

$$\mathrm{RFI}(n + m) = \left[\frac{A}{m - \delta} + \frac{B}{(m - \delta)^2} \right] W_m \qquad (10.9)$$

A and B are complex numbers, so now the model for RFI appears to have five variables! Clearly, if (10.9) is the FFT response to the signal modeled by (10.6), there must be one constraint on A and B, but we need not concern ourselves with it. The entire process of modeling and calculating the parameters of the model is rife with assumptions and approximations anyway, so the only valid test will be how well the process cancels actual ham signals.[32]

Calculation of the Parameters. δ can be calculated from $\mathrm{RFI}(n)$ and $\mathrm{RFI}(n + 1)$ [$m = 0$ and 1 in (10.6)] if the B term is ignored:

$$\frac{|\mathrm{RFI}(n)|}{|\mathrm{RFI}(n + 1)|} = \frac{1 - \delta}{\delta} \frac{W_1}{W_0} \qquad (10.10)$$

so

$$\delta = \frac{|\mathrm{RFI}(n + 1)/W_1|}{|\mathrm{RFI}(n + 1)/W_1| + |\mathrm{RFI}(n)/W_0|} \qquad (10.11)$$

This equation requires two operations that are very tedious in DSP, square-root and division, so an iterative Newton procedure using $|\mathrm{RFI}(n)|^2$ and $|\mathrm{RFI}(n + 1)|^2$ must be used. Then A and B are given by

$$A = \frac{-\delta^2 \mathrm{RFI}(n)}{W_0} + \frac{(1 - \delta)^2 \mathrm{RFI}(n + 1)}{W_1} \qquad (10.12)$$

$$B = \delta(1 - \delta) \left[\frac{\delta \mathrm{RFI}(n)}{W_0} + \frac{(1 - \delta)\mathrm{RFI}(n + 1)}{W_1} \right] \qquad (10.13)$$

Estimation and Subtraction of the RFI on the Tones Used. Evaluation of (10.9) for $m < 0$ and $m > 1$ requires "clever programming tricks" to avoid divisions and minimize RAM storage; these are at the moment proprietary. It was reported in [RFI2] that if the ham interference is simulated by random noise that is band limited to 2.5 kHz and SSB modulated, as prescribed in an earlier version of the VDSL system requirements, then better than 60 dB of cancellation can be achieved across the full band; results in [RFI4] appear to confirm this.

---

[32] The recipe may be very complicated and the cook may make several changes, but the proof of the pudding is still in the eating!

I have seen no reports of results with the more complicated ham model defined in [Cioffi, 1998].

10.6.5 Unfinished Business

Ham Ingress Seen at the Other End. The worst-case differential-mode interference *added to* the line signal is at 0 dBm, and this will propagate in both directions. This is small compared to the 11.5 dBm total upstream transmitted power, but it is very narrow band. If the ham band is halfway between two tones, it can be considered to be spread over $2\Delta f =86$ kHz, and to have a PSD of approximately -50 dBm/Hz: 10 dB higher than most upstream tones. It is unlikely that front-end commom-mode/differential-mode analog cancellation can be used, so a combination of digital cancellation and FFT sidelobe suppression adding up to at least 10 dB more than the SNR expected on those tones will be needed.

Cancellation of AM Radio. The situation with AM radio is different from ham radio in that:

1. The levels of ingress are typically lower. [Cioffi, 1998] specifies that at a distance of 300 ft from an AM radio transmitter the differential-mode interference level is -30 dBm (compared to a worst-case 0 dBm for ham radio). This is not low enough, however, that cancellation is not needed, only that it need not be so good.
2. The bandwidth is much wider: typically, 10 kHz compared to 4 kHz.
3. The signal is present (nearly) all the time, and the center frequency is constant.
4. Because there is no egress limitation, there is no PMD-layer prohibition of transmission in the AM bands.[33] Therefore, the silencing of tones to allow estimation of the interference must be handled by system management. It might appear that these tones would be silenced during the normal handshake process just because their SNR was too low, but until some cancellation is done, many more than two tones would appear to be unusable. Cancellation must be done before bit loading.

I know of no algorithms for dealing specifically with AM interference.

10.7 COMPARISON AMONG FDD, ZIPPER, AND SDMT

The three systems can be compared under 12 different criteria. Under each we consider the systems both in general and in the specific implementation proposed. There are two types of criteria here: those that are fundamental to the

---

[33] There are too many of them for this to be feasible anyway.

systems, for which performance or implementation can be quantified without much controversy, and those for which a system's presently perceived disadvantage may decrease or even disappear after further development and demonstration.

A pure FDD DMT system has not been proposed for VDSL, but it is useful to define one as a benchmark. A 512-pt FFT with a cyclic prefix of 40—just as for ADSL and SDMT—would seem like a reasonable place to start. For many of the criteria concerned with data rate the performance of such a system would be slightly worse than Zipper. Both suffer (Zipper slightly less so because of its lower sidelobes and sharper band edges) from the inflexibility that comes from the difficulty of building a digitally programmable analog receive filter. Both therefore fall in the "may improve later" category.

10.7.1 Efficiency

In theory, and with equally flexible choice of symbol and superframe durations, SDMT and Zipper could be equally efficient: For both of them the overhead (inefficiency) depends on the sum of the cyclic prefix, which is determined by the IR of the channel, and the guard period (cyclic suffix), which is determined by the propagation delay of the loop. In practice, general-purpose and ISDN-compatible SDMT achieve 41.7% and 45.4%, respectively, compared to Zipper's 46.4%; FDD, with typical 15% guard bands, is around 42%.

10.7.2 Latency

Because Zipper and FDD do not have to "turn the line around" and do not rely for their efficiency on "long" transmission blocks in each direction, their latency is lower than SDMTs. The end-to-end processing delay of a DMT system, however, is typically four symbols, so for Zipper, with its eight-times-longer symbol, much of this improvement is lost.

10.7.3 Mixture of Symmetric and Asymmetric Services and Coexistence with ADSL

FDD and asynchronous Zipper, both of which are forced to use large contiguous blocks of tones, are equally inefficient in assigning bands for mixed (i.e., including ADSL) services.[34]

For TDD-compatible applications, all ADSL and VDSL systems must be synchronized to the ISDN. Therefore, they must be synchronized to each other, and synchronous Zipper can be used. Zipper and SDMT would be equally flexible in dealing with mixed services.

---

[34] General-purpose SDMT would be much more efficient than either, but I promised I would not talk about that!

10.7.4 RFI Egress Control

Because Zipper uses an FFT that is eight times larger than for FDD and SDMT, the subcarrier spacing is one-eighth, and the sidelobes of the transmit IFFT fall off eight times faster. This means that just turning off tones in the "forbidden" ham bands may be sufficient to supress the PSD by the required 20 dB; FDD and SDMT have to do a little extra processing to achieve this. If in some installations—depending on cabling practices (burying, shielding, twisting, etc.) of each LEC—egress control is considered more important upstream than downstream, then Zipper could reserve tones in the ham bands for downstream only.

10.7.5 Analog RFI Cancellation

Analog RFI cancellation might appear to be independent of duplexing and modulation, but it is not entirely. An important part of the common-mode-to-differential-mode canceler described in Section 10.3.1 is the periods without a transmit signal that are used for adaptation; fast adaptation of the taps in the presence of such a signal would be much more difficult. FDD and Zipper do not include any quiet periods, and no results with a canceler have been reported.

10.7.6 Digital RFI Cancellation

The cancellation methods described in Section 10.3.3 would appear to be almost equally applicable to any DMT system. The lower sidelobes of Zipper would be an advantage, but its longer symbol period means that the modeling of the ham signal has to be more precise.

10.7.7 AFE Performance

Zipper's need to reduce the AFE gain by anything from 15 to 25 dB to deal with the reflected transmit signal will greatly increase the importance of the noise floor.

10.7.8 Complexity: AFE and ADC

From the calculations in Section 10.5.4 and the wideband RL of approximately 6 dB shown in Figure 10.9(b), it appears that a Zipper system needs about 4.5 extra bits in its ADC to achieve the same SQNR as FDD and SDMT.

10.7.9 Complexity: FFTs

Zipper uses two 4096-pt FFTs compared to one 512-pt for SDMT and two 512-pt FFTs for FDD. Depending on the method of implementation—ASIC or DSP—this would be either a serious or very serious disadvantage.

10.7.10 Complexity: Equalizer

The narrower subchannels, lower sidelobes, and adjustable cyclic prefix of Zipper will reduce the effects of line distortion and simplify the equalizer, but I know of no analysis of this effect.

10.7.11 Complexity: Bit Loading Algorithm

The wide variation of subchannel capacity across the band compared with constancy across the symbols means that an optimal partitioning (between down and up) and bit loading will be more complex for FDD and Zipper.

10.7.12 Power Consumption

The importance of total power consumption in a VTU-C or VTU-R will depend much on the enclosure used (set-top box or computer for VTU-R, sparse or dense rack for VTU-C), but there is complete agreement that it is very important in a VTU-O; let us consider that.

Analog. The three biggest analog power consumers are line drivers, filters, and ADCs, probably in that descending order.

- An SDMT line driver for a symmetrical system, for example, is on only half the time, but during that time for the same performance it must transmit twice the power; therefore, the line driver powers are about equal.
- SDMT needs almost no filters; FDD needs only one fairly complex one if the rest of the filtering is done digitally; Zipper/DD needs either one fairly simple receive filter or four more bits in its ADC.
- SDMT can turn off its ADC during transmission (half of the time for a symmetrical service, eight-ninths of the time for asymmetrical) and may need many fewer bits.

Digital. Two large FFTs will certainly require more power than one or two medium FFTs. This is a very imprecise statement, but it is difficult to say much more. Power consumption depends on rapidly evolving VLSI technology, and estimates of relative wattages have been widely varying, highly partisan, and probably ephemeral. As VLSI technology develops, the importance of the digital power relative to the analog power will decrease.

10.7.13 Synchronization

For the TDD-compatible application synchronization of all the superframes in a binder group has been demonstrated for SDMT. It is still an unknown for Zipper.

TABLE 10.7   Relative Merits of FDD and Zipper for General-Purpose VDSL, and SDMT and Zipper for TDD-Compatible VDSL

	General-Purpose VDSL		TDD-Compatible VDSL	
Criterion	FDD	Zipper	SDMT	Zipper
Efficiency	−1	0	0	0
Latency	+1	0	−1	0
Mixed services (including ADSL)	−3	−2	0	0
RFI egress control	0	+1	0	+2
Analog RFI cancellation	0	0	+2	0
Digital RFI cancellation	0	−1?	0	−1?
AFE performance	0	−3	0	−3
Complexity				
AFE and ADC	−2	−6	0	−6
FFTs	0	−5	0	−5
Equalizer	−3	0	−2 → −1	0
Algorithms	−1	−1	0	−1
Power consumption				
Analog	−2	−3 or −1	0	−3 or −1
Digital	0	−5 → −2	+1	−5 → −2
Synchronization	n/a	n/a	0	−2?

10.7.14   Summary

The estimated relative advantages of the three systems are shown in Table 10.7; the scale is -10 (a fatal flaw) to $+10$ (a triumph), with a zero approximately in the middle. A " $\rightarrow$ " and a second number show my estimate as to how the rating will change (always improve) with further analysis and development.

10.8   A LAST-MINUTE PERSONAL FOOTNOTE

In the spring of 1999 it seemed likely that Zipper or some form of FDD would be chosen (or rather that TDD will be eschewed) for VDSL because the LECs do not have a method of synchronizing CLEC VTU-Cs and multiple VTU-Os. Therefore, I have explained Zipper as well as I can. I even forbore from including SDMT as a candidate for general-purpose VDSL in the comparison in Table 10.7. So, having done my duty, I can now indulge myself and be either a visionary or a cantankerous crank (depending on your point of view). A few years from now you can read this as history and say "So what?", "If only!", or perhaps even (being very optimistic) "Thank goodness!"

10.8.1   Duplexing

In the fall of 1999 it now appears that FDD is the favored method of duplexing; let us consider this.

If I had considered SDMT as a candidate for general-purpose VDSL it would have surpassed Zipper under all criteria except two: RFI egress control (a very small -1) and synchronization (-6 improving to -3). Zipper will always be more complex (the RAM size is immutable), consume much more power, and be inferior in performance; furthermore, no system that meets worldwide requirements has been demonstrated.[35] FDD advocates will counter by saying that no SDMT system to synchronize CLECs or multiple VTU-Os has been demonstrated either. True! So nobody has a universally workable system, and *a standard would be premature.*

The recommendation by the FSAN group [FSAN, 1999] that TDD not be used is foolishly shortsighted and may force the adoption of an expensive and suboptimal solution. It is interesting to consider how this happened. In 1995 it was decided that a VDSL transceiver should be field-configurable for symmetric or asymmetric operation on a wide range of loops. Amati and Telia realized that the filters for conventional FDD would have to be switchable; there was no practical way to do this, so SDMT and Zipper were developed as unconventional solutions to the problem. Unfortunately, however, both companies greatly underestimated the importance of their respective problems—synchronization and the ADC, respectively—and the acrimony of intellectual and commercial competition prevented any serious attempts to solve those problems.

Since it is now clear that VDSL that coexists with TDD ISDN will have to solve *all*[36] the synchronization problems, and that most of them have already been solved in Japan in order to make the TDD system work, it would make sense for there to be one comprehensive and coherent standard for VDSL.

10.8.2 Modulation

It also appears that CAP is the preferred modulation method because no DMT FDD system has been proposed, even though it must be obvious to all that an "ADSL-like" DMT FDD system could be easily built, and clear to many that it would be superior in such matters as egress control and RFI cancellation. This produces a bizarre situation and a sad commentary on the standardization process. The selection of a system as a standard should, ideally, be based on its intrinsic merits, not on the efficiency of a particular implementation, yet T1E1 appears ready to reverse itself from ADSL to VDSL!

---

[35] All tests appear to have been done without bridge taps and with THLs >25 dB. I know of no 14-bit ADCs at sampling rate above 22 MHz that would be needed to deal with RLs as low as 5 dB.

[36] There will almost certainly be ILECs in Japan with separated "COs."

11

FUTURE IMPROVEMENTS

Until now we have discussed mostly what can be done and what has been done. In, this chapter we discuss what *might* be done. If some of the ideas are proven to be useful, it should be possible to incorporate them seamlessly[1] into G.992 and G.VDSL via the option exchange defined in G.994 (handshake protocol for DSL modems).

CAVEAT: The reader is advised that most of the ideas in this chapter have, for several reasons—lack of time being the main one—not been subjected to peer review. They are offered therefore only as food for thought, with no claims made as to their nutritional value.

11.1 FREQUENCY-DOMAIN PARTIAL RESPONSE

In Section 7.1.1 we saw how frequency-domain partial response (FDPR) with a $(1 - \Delta)$ correlation as defined by equation (7.1) reduces the sidelobes of the associated (I)FFT; it may also have other useful effects, depending on whether it is introduced in the transmitter or the receiver.

11.1.1 FDPR in the Transmitter

With a $(1 - \Delta)$ correlation, each symbol of the transmit signal has a sinusoidal envelope as shown by equation (7.3) and has much reduced sidelobes as shown in Figure 7.2. It can be used with a quiet guard period (not a cyclic prefix) of ν samples, and just as for an unshaped envelope, only IR terms beyond the range of the guard period (i.e., h_i for $i > \nu$) contribute to the distortion.

Effects of Distortion. The calculations in Section 6.2.1 and particularly equation (6.11) are based on the assumption that all transmit samples have equal average energy. If, however, the envelope is shaped as shown by (7.3), the same

---

[1] See [Krechmer, 1996] for a discussion of etiquettes and protocols to accommodate innovations.

method of analysis can be used to show that the general form of (6.11) is modified to

$$|H_0|^2 + |H_1|^2 = 2\sum_{i=1}^{N} c_i^2 h_i^2 \tag{11.1}$$

where, as before, $c_i = 0$ for $i \leqslant \nu$ but now

$$c_i = \sum_{m=1}^{i} \sin^2\left(\frac{\pi m}{N}\right) \qquad \text{for} \qquad N > i > \nu \tag{11.2}$$

The total signal energy of each symbol is modified from (6.10) to

$$|H|^2 = c_N \sum_{i=0}^{N-1} h_i^2 = N_{\text{car}} \sum_{i=0}^{N-1} h_i^2 \tag{11.3}$$

Hence

$$\text{STDR} = N_{\text{car}} \sum_{i=0}^{N-1} h_i^2 \bigg/ 2\sum_{i=0}^{N-1} c_i h_i^2 \tag{11.4}$$

Equation (11.4) represents a big increase in STDR from (6.13), but the question of whether the SDRs on all the subcarriers are reduced proportionally is *unfinished business*.

11.1.2 FDPR in the Receiver

The benefits of reduced sidelobes in the receiver are reduced noise enhancement and greater RFI cancellation. A disadvantage is that both the received signal and the noise are $(1 - \Delta)$ correlated, and the lost 3 dB relative to AWGN cannot be retrieved. Considering that it *should* be possible to control noise enhancement by careful equalizer design, and that the RFI canceler described in Section 10.6.3 works well enough without the correlation, it would seem that FDPR should not be used in the receiver.[2]

11.1.3 Filterless FDD

An ATU-C transmitter using FDPR could zero all tones in band 1 (for G.992 Annex A operation) or in bands 1 and 2 (for Annex B), and the sidelobe

---

[2] A fixed FDPR using $(1 - \Delta)$, that is; an adaptive correlation using more taps is discussed in Section 11.6.

attenuation would be such that it would need almost no transmit filter. Most ATU-R transmitters will use 64 tones to allow for configuration as a G.992 Annex B modem, so when operating as an Annex A they will be able to zero tones in band 2. Unfortunately, receive filters seem to be unavoidable! Even if FDPR were used in the receiver, the in-band part of the transmit signal would still demand more bits in the ADC unless removed by an analog filter.

11.1.4 Unfinished Business: Coding for FDPR to Retrieve "Lost" 3 dB

[Nasiri-Kenari et al., 1995] described coding for a time-domain PR channel. It may be possible to use the same approach for frequency-domain PR.

11.2 EQUALIZATION

In Section 8.4.4 we described an equalizer such as is used in first-generation ADSL systems and mentioned some of the problems. Very few details of the performance of such an equalizer are available, but it is known that particularly for the upstream signal in an FDD ADSL system, unequalized distortion is the dominant impairment (i.e., greater than noise) on some subcarriers, resulting in significant reduction in capacity. There are clearly many problems remaining to be solved; in the next two subsections we define the problems and suggest some possible solutions.

11.2.1 TEQ

The basic TEQ problem is to shorten the channel impulse response from (theoretically) infinite to $(\nu + 1)$ terms. Even if the IR is effectively finite ($h_i \approx 0$ for $i > nh$ but $nh > \nu$), this shortening cannot be done exactly if W is a one-path FIR, so the error implied in (6.11) can only be minimized. That is, if the equalized IR is defined as $H'(D) = h'_0 + h'_1 D + \cdots, h'_{nh+nw} D^{nh+nw}$ with

$$h'_k = \sum_{i=0}^{k} w_{k-i} h_i \qquad (11.5)$$

and a delay of d samples is allowed for, then

$$E \approx \left[\sum_{k=0}^{d-1} (d-k) h_k'^2 + \sum_{k=d+\nu+1}^{nh+nw} (k-d-\nu) h_k'^2 \right] \Big/ \sum_{k=d}^{d+\nu} h_k'^2 \qquad (11.6)$$

should be minimized. Furthermore, the dynamic range of W must be constrained so that (6.17), or some slightly relaxed version of it, is satisfied.

NOTE: H has four poles that are caused by the two second-order high-pass filters, and these are known within the tolerance of the LC components; it is very tempting to include four canceling zeros in W as explicit fixed factors or in a starting point for an iteration. If this is done, however, then in the two-octave-wide ADSL upstream band from about 30 to 120 kHz, W may have a dynamic range of 48 dB. If the dominant crosstalk were kindred NEXT, which increases 9 dB across that same band, the potential for noise enhancement would be very serious. B must have other zeros to reduce the variation of W across the band.

A Conjecture. All simulation results that I have seen show SDR(f) almost monotonically increasing with frequency (see, e.g., Figure 4 of [Pal et al., 1998]); the SDR is often lower than the SNR at low frequencies but very much greater (wastefully so?) at high frequencies. I believe, however, that these SDRs all resulted from minimizing the equally weighted error (the "flat wall"). If the "tapered wall" of (6.11) were used, greater weight would be given to the more delayed IR terms. Since these correspond to the low-frequency components of the remanent distortion, the SDR at low frequencies should be increased.

Polypath Implementation

NOTE: This section was written after much correspondence and conversation with an ex-colleague, Debajyoti Pal. The ideas are his; only the words are mine.

One improvement on the structure of a TEQ was proposed in [Pal et al., 1998]. As shown in Figure 11.1, the receive signal is split into two or more paths by oversampling. Each of the sampled IRs H_a, H_b, ... is acted upon separately by a subequalizer with coefficients $w_{a,i}$, $w_{b,i}$, ..., and the outputs are added. For the two-path case the requirement is

$$H_a(D)W_a(D) + H_b(D)W_b(D) = D^d B(D) \qquad (11.7)$$

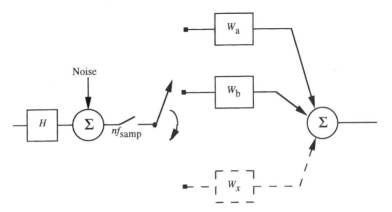

Figure 11.1 Polypath equalizer.

where D is the delay operator ($= z^{-1}$), and d is some to-be-determined delay. If, as before, H_a and H_b are considered to be limited to nh terms, and $B(D)$, the SIR, is limited to ($\nu + 1$) terms, then by a generalization of the Bezout identity (see [Kailath, 1980]), a unique solution for the coefficients of W_a and W_b exists if

$$nh + nw + 1 = 2(nw + 1) = d + \nu + 1 \tag{11.8}$$

That is,

$$nw = nh - 1 \tag{11.9}$$

$$d = 2nh - \nu - 1 \tag{11.10}$$

The important point to note is that this solution is exact.

Simulation results show some improvement in SDR over the single-path method for the same total number of coefficients (FIR taps). The method reported in [Pal et al., 1998] is "off-line"—that is, it requires the inversion of matrices—but an on-line method suitable for implementation in a receiver has reportedly[3] been developed.

Remaining Questions. It should be noted that "polypath" is only a better *structure* for an equalizer. It still has many of the same problems that the single-path equalizer has:

- Should the *B* be a DIR or an SIR, that is, predefined or incidentally found during adaptation?
- If it should be predefined, what are the criteria? How much performance might be lost by a poor choice?
- If it should be found by adaptation, how can one ensure convergence to a global minimum?

There is an obvious similarity between *polypath* equalizers and fractionally spaced equalizers [Qureshi, 1985], but Pal appears to have been the first to point out the possibility of an exact solution if *H* is of finite duration. This, however, is a big "if"; how far the superiority of the polypath approach extends when *H* is IIR is *unfinished business*.

11.2.2 FEQ

Channel equalization in the frequency domain was considered in [Jablon, 1989]. This was for SCM, for which all operations had traditionally been performed in the time domain, yet Jablon showed that for severely distorted channels, which required very complex equalizers, there were advantages—from reduced

---

[3] D. Pal, private conversation.

complexity—to transforming into the frequency domain and equalizing there. For MCM signals, which must be in the frequency domain for detection anyway, it would seem that the advantages might be even greater.

The IR of any channel is theoretically infinite in duration, and particularly for some wireless systems with severe multipath problems, it is possible for a significant part of the IR to span many symbols. For DMT xDSL systems, however, the significant part of an IR is usually considerably shorter than one symbol length. Furthermore, it is (see Section 8.4.3) fairly easy (and only very slightly suboptimum) to adjust the symbol timing so as to limit the effect of the IR—operating on the discontinuities at the beginning and end of a symbol—to just the present and the following symbol. With these restrictions the distortion will, as shown in Section 6.1, be split into two identical halves: ICI and ISCI.

At the output of the DFT the received signal vector[4] $\mathbf{y}_k$ for symbol k is related to the present and previous transmit signal vectors, $\mathbf{x_k}$ and $\mathbf{x}_k - 1$, by

$$\mathbf{y}_k = \mathbf{D} \cdot \mathbf{x}_k - \mathbf{M} \cdot \mathbf{x}_k + \mathbf{M} \cdot \mathbf{x}_{k-1} \qquad (11.11)$$

where the diagonal matrix $\mathbf{D}$ is the transform of the circulant matrix of (6.8) and $\mathbf{M}$ is the transform of each of the distortion matrices. $\mathbf{M}$ is theoretically a full $N \times N$ matrix, but in practice it can be limited to the main diagonal and some number of off-diagonal terms on either side. The number of significant off-diagonal terms depends on the amount of distortion and (surprise!) the rate of fall-off of the sidelobes.

The most obvious way to equalize $\mathbf{y}_k$ is shown in Figure 11.2(a). This form is reminiscent of an FFE/DFE structure for a single-carrier system (except that the operations are performed in a block mode in the frequency domain), but it does not seem to have any merits otherwise. Because the feedforward and feedback transfer functions share a common factor, $(\mathbf{D} - \mathbf{M})^{-1}$, Figure 11.2(a) can be simplified to Figure 11.2(b). It is informative to note that if $\mathbf{M} = \mathbf{0}$, which is the condition that is assumed when a guard period and a TEQ are used, then both figures reduce to the conventional $\mathbf{D}^{-1}$ with no feedback. Figure 11.2(a) and (b) both require multiplication by two potentially full matrices, but Figure 11.2(b) can be further simplified to 11.2(c), which requires multiplication by only one full matrix and one diagonal matrix.

Noise Enhancement of an FEQ.   All the structures of Figure 11.2 implement a partial DFE; they remove the effects of half of the distortion (the ISCI) without causing any noise enhancement, and leave only the ICI to be linearly equalized. The overall noise enhancement should therefore be half that of a fully linear TEQ, but what "half" means in this context is not clear.

Figure 11.3 shows an attempt to develop a full MCM DFE from Figure 11.2(b). This is the frequency-domain equivalent of an FFE/DFE combination in the time domain. The matrix $(\mathbf{D} - \mathbf{M})^{-1}$ has been replaced by a "feedforward"

---

[4] Ignoring noise for the moment.

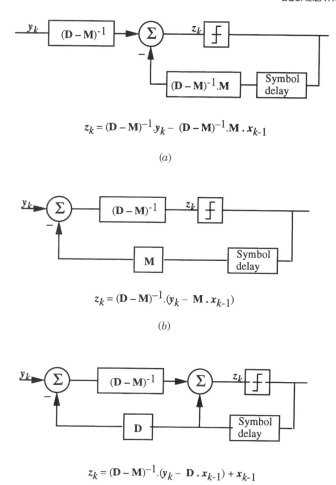

$$z_k = (\mathbf{D} - \mathbf{M})^{-1} . y_k - (\mathbf{D} - \mathbf{M})^{-1} . \mathbf{M} . x_{k-1}$$

(a)

$$z_k = (\mathbf{D} - \mathbf{M})^{-1} . (y_k - \mathbf{M} . x_{k-1})$$

(b)

$$z_k = (\mathbf{D} - \mathbf{M})^{-1} . (y_k - \mathbf{D} . x_{k-1}) + x_{k-1}$$

(c)

Figure 11.2 Three forms of FEQ.

matrix **FF**, which ensures that each subcarrier is distorted only by the subcarriers below it, and an upper triangular "feedback" matrix **FB**. The equivalence is

$$\mathbf{FF} = (\mathbf{D} - \mathbf{M})^{-1}(\mathbf{FB} + \mathbf{I}) \tag{11.12}$$

How the overall noise enhancements of this and the partial DFE in Figure 11.2(b) compare to that of a linear TEQ is *unfinished business*.

Adaptation of an FEQ. An FEQ has the same basic problem for adaptation as a TEQ: that of simultaneously adapting the two sets of parameters $(\mathbf{D} - \mathbf{M})^{-1}$ and **M**, which roughly correspond to the W and B of Section 8.4.4.

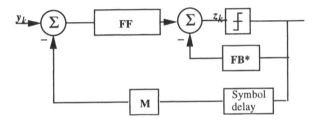

* Note: FB is upper diagonal

Figure 11.3 Full MCM DFE.

11.2.3 TEQ or FEQ?

Performance Optimization. A disadvantage of TEQs is that it is very difficult to relate the performance of the equalized system to any time-domain parameter that can be minimized.[5] The weighted squared error seems to be the best parameter, but as we have seen, the error may affect the performance of the subcarriers in an unequal way that is very difficult to allow for; even the minimum mean square error (mmse) solution that is optimum in the time domain may be far from optimum in the frequency domain.

A potential advantage of FEQs in this respect is that each component of the error vector is very closely related to the error rate on a subcarrier. A weighting factor for each component that is proportional to the SNR on that subcarrier should ensure that the mmse (now summed over the elements of the vector $\mathbf{e}$) is a very good measure of the overall error rate.

Computational Complexity. Depending on the number of off-diagonal terms that are needed in $(\mathbf{D} - \mathbf{M})^{-1}$, the computational requirements for an FEQ are comparable to those for a TEQ. For an FEQ with N(Ftap) coefficients off the main diagonal the number of complex multiplications is $(f_S/2) \times N(\text{Ftap})$, so the number of real multiplications is $2 \times f_S \times N(\text{Ftap})$. By comparison, for a TEQ with N(Ttap) taps the number of real multiplications is $f_S \times N(\text{Ttap})$. The parameters for comparison are therefore N(Ttap) and $2 \times N$(Ftap): probably not a strong influence either way.

Memory Requirements. The big disadvantage of an FEQ, however, is the memory requirement for the coefficients.[6] This is so large that for DSL an FEQ may not be practical in the near future.

---

[5] This is in contrast to a single-carrier system, in which the error rate depends very closely on the mean square error.

[6] Depending on the number of off-diagonal terms in $(\mathbf{D}-\mathbf{M})^{-1}$ and $\mathbf{M}$, the amount of storage required for Figure 11.2(*c*) may be only slightly more than half that required for Figure 11.2(*b*).

Unfinished Business. Another disadvantage of an FEQ that relies on SNR weights to guide its training is that these must be learned (via measurement and calculation) before the equalizer is trained, and then the STNR must be learned again to calculate the bit loading. Whether this sequence can be fitted into the training sequence defined in G.992, for example, remains to be seen.

11.3 ECHO CANCELLATION

I passed on "conventional" ECs in Chapter 8 because they have been well described elsewhere,[7] and I consider the benefits that come from simultaneous transmission and reception on some loops in some small part of the band not to be worth the algorithmic and implementational complexity. In this chapter, however, it is worthwhile to explore ways in which echo canceling could be simplified.

First, however, we must consider the basic conditions under which ECs must operate:

1. One of the results of loop timing (see Section 8.4.2) is that the ATU-R transmitter and receiver are symbol synchronized so, theoretically at least, EC could be performed in the frequency domain after just one FFT performed on the sum of receive signal and echo.[8]

2. At the CO, however, the situation is different. The round-trip delay will not usually be an integer multiple of the symbol period, and transmit and receive symbols cannot be aligned. This makes subtraction in the frequency domain much more complicated.

3. The return loss of the 4W/2W hybrid at both ends will usually be much less than the attenuation of the loop (both positive dB), so either (a) the ADC must convert the extra bits to accommodate the reflected transmit signal,[8] or (b) a precanceler or an adaptive hybrid must be used. The rationale for a precanceler is that bits in a DAC are cheaper than in an ADC; the rationale for an adaptive hybrid is that no extra converter is needed at all. I believe that the adaptive hybrid offers the most promise for future development.

4. The "A" in ADSL means that downstream will use a wider bandwidth than upstream (at least twice as wide, usually eight times), with all the complications of down- and up-sampling in the canceler.

The *potentially* simplest RT EC is shown in Figure 11.4. The correlation and calculation of the echo-emulating coefficients and the subtraction are done in the frequency domain. The main problem with both emulation and subtraction in the frequency domain until now has been that because of the high level of FFT

---

[7] See the specialized bibliography.
[8] See Section 10.4 on Zipper.

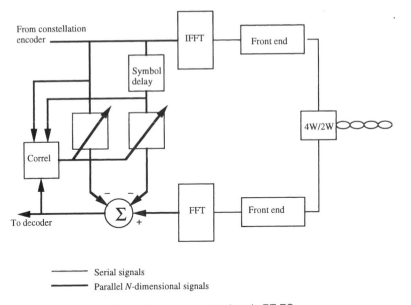

Serial signals
Parallel N-dimensional signals

Figure 11.4 Frequency-domain RT EC.

sidelobes, the number of off-main-diagonal terms in the matrix input to the echo-emulating FFT (and hence both size of RAM and number of multiplications) is prohibitively large. If, however, as suggested in Section 11.1.1, FDPR coding were used in the transmitter, the sidelobes of the transmit signal would be low enough that the number of off-diagonal terms would drop considerably—probably to two or three—and frequency-domain operations might become feasible.

A CO EC must deal with the nonalignment of transmit and receive symbols. If (and it is a big "if") all-frequency-domain operation is feasible for the RT EC, then *perhaps*[9] it can be extended to the CO by performing matrix emulations and subtractions for both the present and previous symbols.

Despite all the efforts to reduce the PAR of the transmitted signal that were described in Section 8.2.11, occasional clips will occur. In "conventional" or "first-generation" ECs, these are dealt with by clipping digitally and passing the same clipped signal to both the transmit DAC and the echo-emulating FIR. With the configuration of Figure 11.4, this could be done only by passing the clipped signal back through an FFT to generate a "reconsidered" frequency-domain transmit signal, a tedious process that would probably completely cancel the benefits of frequency-domain processing. It is possible, however, that this reprocessing would not be needed: In a first-generation EC the main effects of clipping would occur in the precanceler, and it must be fed with the true transmit signal; with an adaptive hybrid, however, this occurs automatically, and the echoes of clips will already be attenuated by about 25 dB.

---

[9] Note the caveat at the beginning of this chapter!

11.4   FRONT-END CROSSTALK CANCELLATION

Crosstalk—both near-end and far-end—is like RFI in that it has both common- and differential-mode components, and many attempts to use a similar "correlation/cancelation" technique have undoubtedly been made. The basic theory can be explained very simply. Let the uncorrupted differential-mode signals on pairs A and B be designated A and B. Then, as we have seen in Section 3.5, B couples into both the differential and common modes on pair A via imbalance of the coupling immittances. The corrupted differential- and common-mode signals on pair A, seen between the outside terminals and at the center tap, respectively, can be expressed as

$$S_{dm} = A + \alpha B \qquad (11.13a)$$
$$S_{cm} = \gamma A + \beta B \qquad (11.13b)$$

where α and β ($\ll 1$) are the crosstalk coefficients and γ is the differential mode-to-common mode conversion ratio for pair A. Cancellation is achieved by learning (α/β) and forming

$$S_{dm} - \frac{\alpha}{\beta} S_{cm} = A\left(1 - \frac{\alpha\gamma}{\beta}\right) \qquad (11.14)$$

This is a much harder task than RFI cancellation for several reasons:

1. α and β vary over approximately the same amplitude range. This means that in contrast to the situation with RFI, the canceler must adapt in a condition of approximately 0 dB NSR; the step size would have to be much smaller, and the final tap jitter would be greater.

2. If pair B couples into pair A, so will pairs C, D, and so on, and with completely different coupling coefficients. Correlation/cancellation of crosstalk would appear to be of value only when there is one dominant crosstalker.

3. The interference occurs across the full band rather than in narrow 4-kHz slot(s), and both coupling coefficients, α and β—particularly if they are due to NEXT—vary rapidly and almost randomly across the entire band. Adapting a W to match their quotient accurately would be very difficult and very unlikely to succeed.

After such a pessimistic assessment of its prospects, it must be pointed out that correlation/cancellation has one advantage over digital NEXT cancellation as discussed in the next section: it does not require knowledge of the data transmitted on the interfering pairs and could therefore perhaps be used to cancel NEXT in a remote unit, or—pushing the envelope—to cancel FEXT in either unit. This method of cancellation might be used to reduce FEXT from the one dominant crosstalker in a quadded cable.

11.5 DIGITAL NEXT CANCELLATION

Echo cancellation (EC) and NEXT cancellation (NC) perform similar functions: they protect a receiver from "reflected transmissions" on the same pair and on other pairs, respectively. Their principles of operation are also similar: both subtract an estimate of the echo or NEXT from the received signal, correlate the difference with a known transmitted signal, and adapt so as to drive the correlation to zero.

Because signal/NEXT ratios are very much greater than signal / echo ratios, and only about 10 to 20 dB of cancellation would be needed, a NEXT canceler—for a single interfering pair—would be much simpler than an echo canceler; for a DMT system it could probably operate only in the frequency domain after the FFT. Whether cancellation of alien NEXT (e.g., from a single-carrier modem) using frequency-domain (i.e., FFT) methods would be possible could be the subject for a Ph.D. thesis.

In Section 7.3.3 I suggested that with ADSL ranges being stretched to the maximum, NEXT becomes the dominant impairment, and EC is obsolescent. If, however, NC becomes possible and economically feasible, perhaps that will reinstate EC. It should be noted that digital NC is possible only at the CO, where the transmit data on all the pairs *could*[10] be bussed to each ATU-C. It would not be possible at an ATU-R because the other upstream data are not available, but NEXT is not quite as important for downstream reception because (1) ATU-Rs are typically separated and the NEXT from one to another is attenuated somewhat, and (2) except for T1, there are no other signals above about 300 kHz that can crosstalk.

One possible use for a NEXT canceler would be in a "good neighbor" second-generation ATU-C. It would not transmit in band 1 and so would not interfere with others' upstream reception, but it would be able to deal with others in the binder group that did; that is, it would not need EC, but would need NC.

11.6 CANCELLATION OF RF AND OTHER INTERFERENCE

NOTE: This section is based on correspondence and conversation with an ex-colleague, Brian Wiese; most of the ideas are his.

Interference (a.k.a .noise) can be loosely categorized as:

1. *Wideband.* There is little or no correlation of the noise from one subcarrier band to the next.
2. *Narrowband.* The noise extends over only a few subbands, and anything beyond that is strongly correlated with the "main lobe."
3. *Very narrowband.* The bandwidth of the interference is less than Δf.

---

[10] Presenting a very messy task to the designer of the CO racks!

Kindred FEXT is an extreme example of the first category because the noise has the same bandwidth as the overall signal; we have already discussed possible ways of reducing this. Ham radio is an example of the third category, and in Section 10.6.4 we described a canceler used in a first-generation VDSL modem; the general conclusion there was that it is adequate. In this section we discuss very briefly some possible ways of dealing with the middle category.

FDPR, discussed in Section 7.1, correlated the signals from tone to tone using $(1 - \Delta)$; this can be generalized to

$$C(\Delta) = 1 - \sum_{i=1}^{n_c} c_i \Delta_i \qquad (11.15)$$

and the "symbol-by-symbol" correlating matrix, $\mathbf{C}$, can be found as a generalization of (7.1). If this is applied to the output of the FFT, it generates the frequency-domain equivalent of "trailing echoes," which must be removed by decision feedback, as shown in Figure 11.5(a). It also spreads the noise over n_c subbands, and the margin relative to AWGN for all but the n_c highest subchannels will be reduced by

$$\Delta dB = 10 \log_{10}(1 + \sum c_i^2) \qquad (11.16)$$

The form of (11.15) favored by Wiese has $n_c = 2$ with $c_1 = c_2 = 0.5$ with $\Delta dB = 1.76\,dB$.

This frequency-domain correlation of the FFT output can be thought of in two equivalent ways as shown in Figure 11.5(a) and (b):

(a) It filters the output, reduces the sidelobes,[11] and thereby reduces the amount of narrowband interference appearing in all subbands except the n_c adjacent to the interference.

(b) It predicts the noise in each subband from the calculated noise in n_c "previous" subbands, and subtracts the prediction.

Considering the correlator as a noise predictor offers an interesting possibility. If there is little or no narrowband interference, or for tones far removed from such interference, the loss of margin to AWGN may be more than the gain of margin to interference. Therefore, the correlation coefficients (the elements of $\mathbf{C}$) should be defined by

$$y_i' = y_i - \sum_{j=1}^{n_c} c_{i,j} y_{i-j} \qquad (11.17)$$

---

[11] If $c_l = 1$, the far-out sidelobes decrease as $1/n^2$, compared to $1/n$ for conventional MQASK.

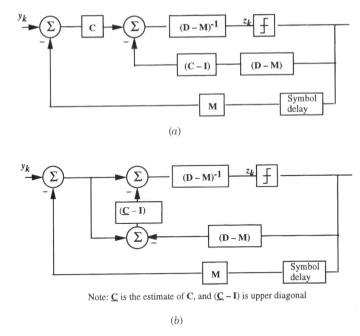

(a)

(b)

Note: $\underline{C}$ is the estimate of C, and $(\underline{C} - I)$ is upper diagonal

Figure 11.5 Frequency-domain correlation: (a) as a noise filter; (b) as a noise predictor.

and be adapted separately for every tone for the best compromise between suppression of interference and increase of noise. It may be sufficient to use just two coefficients per tone, and to constrain them to a maximum of 0.5 each.

11.6.1 Unfinished Business

It has been suggested that the sidelobes of narrowband crosstalkers such as ISDN (both the FDD and TDD versions) could be dealt with in this way; the signals in those frequency regions are certainly strongly correlated with those in the main lobe. Perhaps there are a few Ph.D. students working on this right now.

11.6.2 Grand Finale

These ideas can be incorporated into the FEQ of Section 11.2.4 by changing the linear equalizer $(\mathbf{D} - \mathbf{M})^{-1}$ in Figure 11.5 into a DFE, just as shown in the change from Figure 11.2 to Figure 11.3.

MATLAB PROGRAMS FOR xDSL ANALYSIS

NOTES:

1. For easier reading all comments are *italicized*, and all input statements are **bold**
2. All nested blocks of "for" and "if" statements are progressively indented

A.1 FREQUENCY-DOMAIN ANALYSIS: RESPONSE AND INPUT IMPEDANCES

```
% Calculates response, input impedances and return losses of HPF/loop/HPF;
hold off;
nfdf = input('[number of freqs freq increment] ');
% If nfreq is a power of 2 it will facilitate calculation of IR by IFFT;
nfreq = nfdf(1); df = nfdf(2);
f = df*[0:nfreq – 1];
f(1) = 1e-6; % To avoid dividing by zero;
j = sqrt( – 1); jomega = j*2*pi*f;
% Define unit vector and zero vector;
Uv = ones(1,nfreq); Zv = zeros(1,nfreq);
Rterm = 0.1; Gterm = 10.;

% RL parameters [Rzero, fr, Lzero, Linf, fm, b, C];
% Rows 1 and 2 are for 1 kft of 24 AWG, 26 AWG; (add your own rows to the matrix);
RL = [.0537 .147 .187 .129 .697 .819 15 .7;.0836 .217 .187 .134 .870 .847 15.7];
% Note: C is the same for all gauges of US UTPs; it may not be for other UTPs
khpf = input('With high-pass filter(1), without(0) ');
  if khpf;
  LC = input('[Transformer inductance, Series capacitance] ');
Ahpf = Uv;   Bhpf = Uv./( jomega*LC(2));
Chpf = Uv./(jomega*LC(1)); Dhpf = Uv + Bhpf.*Chpf;
% Note: Source and load highpass filters are assumed the same;
% program could be generalized to allow different filters
% Initialization of M with HPF;
A = Ahpf;   B = Bhpf;
C = Chpf;   D = Dhpf;
```

```
else; % Initialization of M without HPF
A = Uv;  B = Zv;
C = Zv;  D = Uv;
end; % if khp
% Initialization of image attenuation (derived from gammas of in-line sectons);
dBimage=Zv;
nsect = input('Number of sections ');
% Definition of loop, section by section from CO to RT, and build up of A, B, C, and D
for n = 1:nsect;
LGT = input('[Length in kft, Gauge row, Type(1=in-line, 2=b-t)] ');
leng = LGT(1); row = LGT(2); type = LGT(3);
% R per unit length according to equation (3.8);
Rper = RL(row,1)*(Uv + (f/RL(row,2)).^2).^0.25;
% xb = (f/fr)^b;
xb = (f/RL(row,5)).^RL(row,6);
% L per unit length according to equation (3.9);
Lper = (RL(row,3)*Uv + RL(row,4)*xb)./(Uv + xb);
Cper = RL(row,7);
Zseries = jomega.*Lper + Rper;
Yshunt = jomega*Cper;
Z0 = sqrt(Zseries./Yshunt);
gamma = leng*sqrt(Zseries.*Yshunt);
    if LGT(3) = = 1; % In-line section
    An = cosh(gamma);  Bn = Z0.*sinh(gamma);
    Cn = sinh(gamma)./Z0;  Dn = An;
    dBimage = dBimage-8.686*(real(gamma));
    else; % Bridge tap
    An = Uv;  Bn = Zv;
    Cn = tanh(gamma)./Z0;  Dn = Uv; % bridge tap
    end; % if type
Atemp = A.*An + B.*Cn;  B = A.*Bn + B.*Dn;
Ctemp = C.*An + D.*Cn;  D = C.*Bn + D.*Dn;
A = Atemp;
C = Ctemp;
end; % end of loop on n
if khpf; % Add hpf at RT;
Atemp = A.*Dhpf + B.*Chpf;  B = A.*Bhpf + B.*Ahpf;
Ctemp = C.*Dhpf + D.*Chpf;  D = C.*Bhpf + D.*Ahpf;
A = Atemp;
C = Ctemp;
% Note: The load high-pass filter is the mirror image of the source one,
% so Ahpf and Dhpf are interchanged
end; % if khp
% Check that AD-BC = 1; Erase these lines when confidence is established!;
determ = A.*D-B.*C;
plot(f,real(determ),f,imag(determ)); grid on; figure(gcf); pause;

% Response
Hsq = abs(2*Uv./(A + B*Gterm + C*Rterm + D)).^2;
save response nfreq f Hsq totleng;
% Can be used as input to A2.m for capacity calculations
plot(f,10*log10(Hsq)); figure(gcf); grid on; hold on;
plot(f(1:50:nfreq),dBimage(1:50:nfreq),'x');
xlabel ('MHz'); ylabel('Loss dB'); pause; hold off;
```

% Image attenuation is superimposed to show that match is good IF there are no b-ts

% Input impedances and return losses;
ZinCO = (A*Rterm + B)./(C*Rterm + D);
ZinRT = (D*Rterm + B)./(C*Rterm + A);
RLCO = 20*log10(abs((ZinCO + Rterm)./(ZinCO-Rterm)));
RLRT = 20*log10(abs((ZinRT + Rterm)./(ZinRT-Rterm)));
plot(f,RLCO,f,RLRT); grid on; figure(gcf); hold on;
plot(f(nfreq),RLCO(nfreq),'o',f(nfreq),RLRT(nfreq),'x');
xlabel('MHz'); ylabel('Return loss dB');
title('Return losses relative to Rterm: RLCO(o); RLRT(x)'); pause; hold off

A.2   LOOP CAPACITY

Note: This program is a very simple example of the calculation of capacities; the reader is encouraged to develop more useful ones that include VDSL systems, mixes of services and loop lengths, spectral management of transmit PSDs, real roll-off filters, and so on.

% Calculates ADSL capacities with ISDN, HDSL and ADSL crosstalk
% Assumes that response and total length of loop have been calculated
% and stored as "Hsq" and "totleng"
BitCap = input('[Bitcapdown Bitcapup] ');
mardB = input('(margin-coding gain) dB ');
alpha = 0.12*10^(-mardB/10);
f = .0043125*[1:256];
fsq = f.^2;
flpt5 = f.^1.5;
Uv = ones(1,256); Zv = zeros(1,256);
hold off;
% Numbers of crosstalkers;
nXT = input('(nl nh nA)');
nl = nXT(1); nH = nXT(2); nA = nXT (3);

% PSDs and crosstalk
% ISDN BRI
f0 = .08; f3dB = .08;
IPSD = (2*26/80000)*((sin(pi*f/f0)./(pi*f/f0)).^2)./(Uv + (f/f3dB).^4);
IN = (0.8e-5)*IPSD.*flpt5+le-15;
IF = (0.8e-5)*totleng*IPSD.*Hsq.*fsq; *% Probably negligible*
IXT = IN + IF + le-15;
% HDSL
f0 = .392; f3dB = .196;
HPSD = (2*30/(392000))*((sin(pi*f/f0)./(pi*f/f0)).^2)./(Uv + (f/f3dB).^8);
HN = (0.8e-5)*HPSD.*f1pt5;
HF = (0.8e-5)*totleng*HPSD.*Hsq.*fsq;
HXT = HN + HF + 1e-15;
% ADSL;
kEC = input('FDD(0) or EC(1)?');
   if kEC = = 1; nmind = 7; nmaxu = 31;
   else; nmind = 36; nmaxu = 28;
   end;
f3dB = 1.104; *% Downstream has 36 dB per octave roll off above 1.104*

```
APSDd = le-4*[ − Zv(1:nmind-1) Uv(nmind: 256)];
APSDu = 10^( − 3.8)*[Zv(1:6) Uv(7:nmaxu) Zv(nmaxu + 1:256)];
ANd = (0.8e-5)*APSDu.*f1pt5;
Sigd = APSDd.*Hsq;
AFd = (0.8e-5)*totleng*Sigd.*fsq;
AXTd = ANd + AFd + le-15;
ANu = (0.8e-5)*APSDd.*f1pt5;
plot(f,Bd); figure(gcf); grid on; pause;
Rated = .004*sum(Bd(nmind:255));

% Upstream
fup = f(1:32);
Sigu = Sig(1:32);
plot(fup,10*log10(Sigu + le-15)); figure(gcf); grid on; hold on;
    if nl>0;
    plot(fup,10*log10(IXT(1:32)));
    plot(f(10),10*log10(IXT(10)),'mo');
    end;
    if nH>0;
    plot(fup,10*log10(HXT(1:32))'c');
    plot(f(10),10*log10(HXT(10)),'cx');
    end;
plot(fup,10*log10(AWGN(1:32)),'g');
plot(fup,10*log10(AXTu(1:32)),'r');
plot(f(10),10*log10(AXTu(10)),'r*'); pause; hold off;
Noiseu = (nl*IXT(1:32).^a + nH*HXT(1:32).^a + nA*AXTu(1:32).^a).^0.6 + AWGN(1:32);
B = min(log2(1 + alpha*Sigu./Noiseu),BitCap(2));
Bu = round((B>0.5).*B); % Minimum of one bit per with g = + 1.5 dB;
plot(fup,Bu); figure(gcf); grid on; pause;
Rateu = .004*sum(Bu(8:nmaxu));
Sigu = APSDu.*Hsq;
AFu = (0.8e-5)*totleng*Sigu.*fsq;
AXTu = ANu + AFu + 1e-15;
% Signals, noise and capacity;
a = 1/0.6;
AWGN = Uv*10^( − 13.5);

% Downstream
plot(f,10*log10(Sigd + le-15)); figure(gcf); grid on; hold on;
    if nl>0
    plot(f,10*log10(IXT),'m');
    plot(f(50),10*log10(IXT(50)),'mo');
    end
    if nH>0;
    plot(f,10*log10(HXT),'c');
    plot(f(50),10*log10(HXT(50)),'cx');
    end;
plot(f,10*log10(AWGN),'g');
plot(f,10*log10(AXTd),'r'); pause;
plot(f(50),10*log10(AXTd(50)),'r*'); pause; hold off;
Noised = (nl*IXT.^a + nH*(HXT).^a + nA*AXTd.^a).^0.6 + AWGN;
B = min(log2(1 + alpha*Sigd./Noised),BitCap(1));
Bd = round((B>0.5).*B); % Minimum of one bit per with g= + 1.5 dB;
Ratedownup = [Rated Rateu]
```

ORGANIZATIONS, RECOMMENDATIONS, AND STANDARDS

This is a list of all the organizations (with addresses), recommendations, and standards that have been cited in the book. A much longer list with nearly every standard pertaining to DSL is given in [Starr et al., 1999].

B.1 INTERNATIONAL TELECOMMUNICATIONS UNION

Secretariat: helpdesk@itu.ch

Study Group 15 is responsible for xDSL; membership with payment of dues is required for attendance.

Recommendations[1]:

G.703	Primary rate (T1/E1) systems
G.961	ISDN-BRA digital system (DSL line format)
G.991.1	HDSL (first generation: two-pair)
G.991.2	HDSL (reserved for second generation: one-pair)
G.992.1	ADSL: a region-generic main body with three region-specific annexes
G.992.2	"Splitterless" ADSL: colloquially known as G.lite
G.993	Reserved for VDSL
G.994	Handshake protocol for all xDSL modems
G.995	Overview of xDSL recommendations
G.996	Test procedures for xDSL
G.997	Physical layer operations, administration, and maintenance for xDSL

[1]They are just that: recommendations. Compliance is not mandatory.

I.361	ATM layer specification
I.432	BISDN UNI Physical Layer

B.2 AMERICAN NATIONAL STANDARDS INSTITUTE

Information available on Web site at www.t1.org/t1e1

Committee T1E1.4 is responsible for xDSL; meetings are free and open to all.

Standards:

T1.101	Synchronization of digital networks
T1.403	DSI Metallic Interface
T1.413	ADSL (Issue 2, 1998)
T1.601	ISDN Basic Rate, Physical Layer
TR28	HDSL (technical report only, responsibility for standard ceded to ITU)

B.3 EUROPEAN TELECOMMUNICATIONS STANDARDS INSTITUTE

Secretariat: graham.rose@etsi.fr

Committee TM6 is responsible for xDSL; meetings are open only to representatives of European companies.

B.4 ATM FORUM

Information is available on Web site at www.atmforum.com.

B.5 ADSL FORUM

Information is available on Web site at www.adsl.com.

EFFICIENT HARDWARE IMPLEMENTATIONS OF FFT ENGINES

Mitra Nasserbakht

Intel Corporation, E-mail: Mitra.Nasserbakht@intel.com or Bite 5000@aol.com

C.1 OVERVIEW

Efficient algorithms for computing the discrete fourier transform (DFT) have enabled widespread access to Fourier analysis in numerous fields. These application areas span diverse disciplines such as applied mechanics and structural modeling to biomedical engineering. In the signal-processing arena, Fourier theory has been widely used for signal recognition, estimation, and spectral analysis. Fourier analysis has been at the core of many communication system subblocks, such as those used for echo cancellation, filtering, coding, and compression. The ability to compute DFT in realtime and with minimal hardware is the key to the successful implementation of many of these complex systems.

The fast fourier transform (FFT) is an efficient algorithm for computing the DFT of time-domain signals. The focus of this chapter is on the necessary ingredients for the design of FFT processing engines capable of handling data of a real-time nature found in most digital signal processing and telecommunications applications.

This appendix starts with a brief overview of the FFT and its computation. Top-level system requirements, addressing, arithmetic processing, memory subsystem, and data ordering are discussed in Section C.3. Section C.4 is devoted to a discussion of implementation issues for a representative FFT processing engine.

C.2 FAST FOURIER TRANSFORM

In 1807, Joseph Fourier described the Fourier series representation of signals where any periodic signal could be represented by the sum of scaled sine and cosine waveforms. The signal can thus be represented by a sequence of an infinite number of scale factors or coefficients. Given such a sequence, we can also determine the original waveform. For Fourier series representation of finite sequences, we deal exclusively with the discrete Fourier transform (DFT).

At the core of the computation are the following two equations that describe the N-point DFT of a data sequence $x(k)$:

$$X(i) = \sum_{k=0}^{N-1} x(k) W^{ik} \qquad \text{(C.1)}$$

and the inverse DFT of a transform sequence $X(i)$:

$$x(k) = \frac{1}{N} \sum_{i=0}^{N-1} X(i) W^{-ki} \qquad \text{(C.2)}$$

where $\{x(k)\}$ is a vector of N real samples, $\{X(i)\}$ is a N-complex vector with Hermitian symmetry, and $W = \exp[-j(2\pi/N)]$ are called the twiddle factors. The number of operations for computing DFT of a signal using direct DFT method is proportional to N^2.

The invention of FFT is attributed to Cooley and Tukey in 1965. Fast Fourier transforms compute the DFT with greatly reduced number of operations. By adding certain sequences of data after performing multiplication by the same fixed complex multipliers, FFT eliminates redundancies in brute-force DFT computations. This efficiency in computation is achieved at the expense of additional reordering steps to determine the final results. These additional steps, once implemented efficiently, have negligible overhead on the overall compute engine. As a result, FFT is an extremely robust algorithm that lends itself well to machine computation. Since most applications of FFT deal with real-time data sequences, full hardware implementation of the algorithm is desired. Due to the complexity and large amount of hardware required for this type of implementation, many trade-offs need to be considered; these are outlined in the upcoming sections.

For large values of N, computational efficiency can be achieved by breaking the DFT into successively smaller calculations. This can be achieved in the time or frequency domain, as discussed in the following sections. The main observation made by Cooley and Tukey was that when N is not prime, the DFT computation can be decomposed into a number of DFTs of smaller length. This decomposition can be continued until the prime radix for the given value of N is reached. For the case where N is a power of 2, the total number of operations is reduced to on the order of $N \log_2 N$.

C.2.1   Radix-2 FFT Computation

In a radix-2 implementation, the basic building block is a two-point *butterfly* shown in Figure C.1. It has two inputs ($x[0]$, $x[1]$) and two outputs ($X[0]$, $X[1]$)

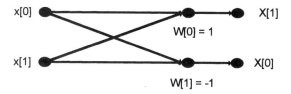

Figure C.1 Basic radix-2 butterfly.

and consists of one complex multiply and two complex adds:

$$
\begin{aligned}
X[1] &= x[0] + x[1] \\
X[0] &= x[0] + W[1] * x[1]
\end{aligned}
\tag{C.3}
$$

In computing the FFT for larger than two-point sequences, data points are successively partitioned according to methods outlined in Sections C.2.3 and C.2.4 until the basic butterfly, shown in Figure C.1, is reached. Figures C.3 and C.4 depict examples of such partitioning. Radix-2 implementations have minimal demand on memory subsystem sophistication but are not the most efficient for large data sequences.

C.2.2 Radix-4 FFT Computation

In a radix-4 implementation, the basic building block is a four-point butterfly, as depicted in Figure C.2. A basic implementation of this requires four additions and three complex multiplication operations:

$$
\begin{aligned}
X[0] &= x[0] + x[1] + x[2] + x[3] \\
X[1] &= (x[0] - jx[1] - x[2] + jx[3])W^1 \\
X[2] &= (x[0] - x[1] + x[2] - x[3])W^2 \\
X[3] &= (x[1] - jx[1] - x[2] - jx[3])W^3
\end{aligned}
\tag{C.4}
$$

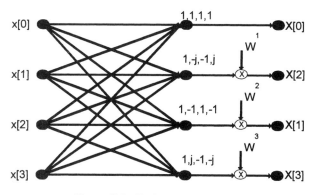

Figure C.2 Basic radix-4 butterfly.

Similar to radix-2 computation, larger values of N need to be divided into groups of 4 this time until the basic butterfly is reached.

C.2.3 Decimation in Time

As mentioned above, it is possible to break up the input sequence to be transformed into successively smaller time-domain sequences, hence the name *decimation in time* (DIT). The DIT can be interchangeably used with decimation in input for the case of a DFT-only engine. In a more general case, input can be the data sequence or its respective transform sequence. For the special case of N being a power of 2, the input sequence is divided into its even and odd parts, and each is so further divided until the base of the algorithm is reached (two points in a radix-2, four points in a radix-4 algorithm, etc.) This requires logr N stages of computation.[1] The total number of complex multiplies and adds will therefore be N logr N.

An example of FFT decomposition is depicted in Figure C.3 for $N = 8$ and a radix of 2. The DIT decomposition is sometimes referred to as the *Cooley–Tukey method* of computing the FFT.

C.2.4 Decimation in Frequency

When performing pure DFT functionality, the compute engine may be designed to divide the frequency domain (output in this case) into smaller sequences. The total number of computations remains the same as for DIT-type algorithms: N logr N, with the number of complex multiplies being (N/radix) logr N and the number of complex additions being N logr N.

In general, a great degree of correspondence exists between the DIT and decimation in frequency (DIF) algorithms whereby interchanging one's outputs and inputs and reversing the direction of the flow graph of computation would yield the other method. In addition, if the input data are entered in normal order, proper permutations must happen to produce bit-reversed ordered data before the final results are returned.

An example of FFT decomposition is depicted in Figure C.4 for $N = 8$ and a radix of 2. The DIF decomposition is sometimes referred to as the *Sande–Tukey method* of computing the FFT.

C.3 ARCHITECTURAL CONSIDERATIONS

In specifying the architecture and design of an FFT engine, there are several key factors to consider. First is the basic FFT computational building block. At a high level, this affects the radix chosen along with memory system availability and requirements. At a lower level, it determines the details of how the radix-x processor engine is to be implemented. These are discussed in Section C.1.1.

---

[1] For convenience, logr is written for $\log_{\text{radix}}$.

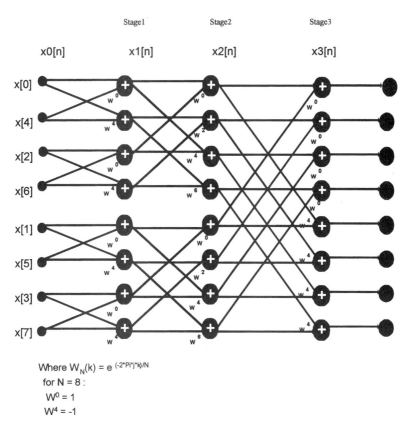

Stage1   Stage2   Stage3

x0[n]   x1[n]   x2[n]   x3[n]

Where $W_N(k) = e^{(-2*Pi*j*k)/N}$
for N = 8 :
$W^0 = 1$
$W^4 = -1$

Figure C.3 Flow graph of FFT: decimation in time (DIT) with $N = 8$.

Another important consideration is memory access. This deals with the number of memory devices to be used, and most important, the addressing scheme used in the implementation of the FFT engine. For space conservation, in-place algorithms are almost universally preferable. In such cases, care must be taken in choosing an appropriate addressing scheme for both data and the corresponding twiddle factors used in different stages of FFT computation, as discussed in Section C.2.

Finally, there are details to consider pertaining to the order in which data and twiddle factors enter the computation engine as well as scrambling and descrambling of data. Some of these considerations are discussed in Sections C.1.1 and C.2.

C.3.1 Number Representation Scheme

One of the most important factors in deciding the architecture of a number-crunching machine is its number representation scheme. This decision directly affects the storage requirements of the machine. In applications such as FFT or

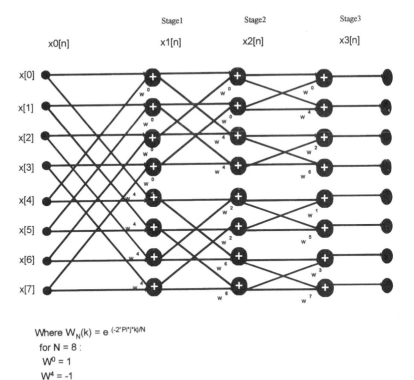

Where $W_N(k) = e^{(-2^r Pi \cdot j \cdot k)/N}$
for $N = 8$:
$W^0 = 1$
$W^4 = -1$

Figure C.4 Flow graph of FFT: decimation In frequency (DIT) with $N = 8$.

other signal processing applications, an increasingly larger amount of on-chip memory is required to support the real-time nature of these applications. Saving a few bits for each data point that needs to be stored in the RAMs will save a large percentage of chip area in such applications.

Twos-Complement Versus Sign-Magnitude. The number representation scheme dictates the type of computational elements used in the architecture. If any data representation system other than the standard twos-complement binary is chosen, one must make a trade-off between taking advantage of the number representation system in computational elements as well, or use standard twos-complement binary arithmetic elements and make appropriate conversions between data formats when necessary. More details of such a trade-off are covered in the Section C.4.

Floating-Point Versus Fixed-Point. The choice of number representation scheme is also driven by the need to achieve maximum "dynamic range" for the available hardware and to minimize error accumulation due to lack of precision. The first floating-point digital filter was built in 1973. Since then several general-purpose DSP machines have been developed to accommodate high-precision calculations. Floating-point signal processing has several advantages over fixed-point arithmetic, the most notable of which is its superior dynamic range behavior. In fixed-point number representation schemes, the dynamic range grows only linearly (6 dB/bit) with increasing bit width, while for floating-point numbers it grows exponentially with increasing exponent bit width. The use of floating-point number representation improves the dynamic range, and it may be the only scheme that provides proper signal/noise ratio (SNR) in the presence of storage limitations.

The improvement in dynamic range comes at the expense of increased hardware complexity. In the case of an FFT engine, this hardware overhead is multiplied by the requirement of having to store initial and final results as well as intermediate values and keeping proper levels of precision.

Due to the inherent round-off noise in computation-intensive applications, the number representation scheme and the carrying of the round-off noise through the computation become essential factors affecting the final performance of the FFT engine. The processor needs to allocate enough internal storage between RAM stages to ensure that the SNR never falls below the minimum system requirements.

In general, the speed of operation for fixed-point multipliers and adders has increased due to technology scaling as well more efficient algorithms. Sub-1nS 64-bit adders have been reported in 0.5-μm technology. Floating-point arithmetic processors have started to enjoy similar performance improvements even without full custom implementations. As a result, system SNR requirements and available hardware are the main driving factors in the implementation of the arithmetic units for FFT engines rather than delay considerations.

C.3.2 Memory Subsystem

One of the more significant considerations in the design of a specialized high-performance DSP engine is its memory subsystem architecture. As the processing engine becomes capable of more speedy operations and as the sophistication of DSP algorithms increases, there is a greater need for fast, efficient memory-access algorithms independent of the advances in DRAM and SRAM technologies. The goal in most such architectures in general and FFT specifically is to satisfy the system bandwidth requirements while occupying the least amount of space. In addition, due to intensive pipelining and the emphasis on high throughput, decoupling of fetch hardware and execute engine is dictated. This allows a continuous stream of data flowing into the compute engine and similarly at the output for further processing and possibly transmission and receipt of data.

In an FFT-specific engine, the basic design of the memory subsystem is influenced by the chosen radix, the number and types of memory elements, and the type of storage. In-place storage is used almost universally to minimize area in applications where it is critical.

FFT Address Storage (FAST). In this section the "FAST" addressing scheme is introduced, which optimizes storage for an in-place FFT algorithm in any radix. Given the number of data points (N) and chosen radix (r), this scheme determines the optimum storage location for each data point in an r-bank memory system.

Addressing for FFT Modes. The N data points can be organized into r banks for a radix-r implementation with N/r data points residing in each data bank. Each data point is assigned a bank number (B), and an address (A) which determine its location within each bank. This assignment needs to be done such that the following conditions are met at all times:

1. Only the locations being read in the current FFT cycle are overwritten.
2. The destination locations are determined such that the data points required for all subsequent butterfly stages reside in different memory banks.

The second requirement poses more restrictions on the address generation hardware. The following algorithm computes the value of bank (B) and address (A) from each data point index (i);

1. Express the index in radix-r notation:

$$i = i(r-1)r^{(r-1)} + \cdots + i(3)r^3 + i(2)r^2 + i(1)r + i(0) \qquad \text{(C.5)}$$

2. Compute the value of the bank (B) according to the following equation:

$$B = (i(r-1) + \ldots + i(3) + i(2) + i(1) + i(0)) \text{ modulo } r \qquad \text{(C.6)}$$

3. Compute the address location (A) for each data point according to

$$A = i \text{ modulo}(N/r) \qquad \text{(C.7)}$$

These equations yield the optimum address and bank assignments for any radix(r), provided that the system is capable of providing r-banks of memory per required storage device. This is sufficient to provide contention-free in-place storage assignments for all FFT stages; there would be logr N stages for a radix-r implementation of FFT. Each stage of the FFT algorithm requires fetching a different set of inputs that will be written back in-place once the computation on each data point is completed. This assignment remains unchanged from stage to stage while preserving the contention-free nature of the algorithm. Table C.1 shows the bank and address assignment for the first nine entries for a 256-point FFT implemented in radix-4, using the assignment scheme above.

Addressing for Pre-/Postprocessing. To reduce computational complexity, the same engine can be used to process FFT and IFFT data. Often, system

TABLE C.1 Memory Assignment for a 256-Point FFT in
Radix-4

Data-Point Index i	Radix-4 Digits $i(3)$ $i(2)$ $i(1)$ $i(0)$	Memory Bank/Address
0	0000	0/0
1	0001	1/1
2	0002	2/2
3	0003	3/3
4	0010	1/4
5	0011	2/5
6	0012	3/6
7	0013	0/7
8	0020	2/8

requirements dictate and greatly benefit from this reduced cost. An example of such a system would be a communication device that is receiving and transmitting information in time-multiplexed fashion. Since the inputs to the FFT and the outputs from the IFFT are "real," a significant reduction in computational resources can be achieved. Instead of zeroing out the imaginary part, we can pre/postprocess the data to enable it to use an engine that is half the size of what would be needed otherwise. Preprocessing is an additional stage before entering the IFFT mode of the engine in which data are prepared. Similarly, postprocessing is an additional stage after the FFT mode of the engine.

Preprocessing modifies the input vector so that when an IFFT is performed the real part of the output vector contains the even time samples and the imaginary part of the output contains the odd time samples. For an effective 512-point IFFT function, the preprocessing function is described by the following equation:

$$Y[i] = \{[X(i) + X^*(256 - i)] + j[X(i) - X^*(256 - i)]W^{(256-i)}\}^* \qquad (C.8)$$

The input to the FFT engine is a 256-point complex vector that is formed from the real 512-point data vector. The real part of the vector is comprised of the even sample points while the odd samples make up the imaginary part of the vector. Postprocessing is done as the final stage according to the following equation:

$$Y[i] = \left(\frac{1}{2}\right)[X(i) + X^*(256 - i)] - (j/2)[X(i) - X^*(256 - i)]W^i \qquad (C.9)$$

These two modes have special requirements in terms of storage and addressing modes.

As can be observed from (C.8) and (C.9) for pre/postprocessing, regardless of the radix used in the computational engine, only two data points are operated on at any instant. This is in line with the radix-2 FFT as only two inputs are fetched and operated on at any given time. In order to reduce the number of cycles

TABLE C.2 Data Ordering/Digit Reversal in Radix-2 and Radix-4 FFT

Data Point	Radix-2 Input	Radix-2 Output	Radix-4 Input	Radix-4 Output
0	0000	0000	00	00
1	0001	1000	01	10
2	0010	0100	02	20
3	0011	1100	03	30
4	0100	0010	10	01
5	0101	1010	11	11
6	0110	0110	12	21
7	0111	1110	13	31

required to perform these two operations and to take full advantage of existing hardware in cases of non-radix-2 FFT architecture, the addressing scheme is modified to pull and operate on "r" data-points. For a multiple of 2 such as a radix-4 case, this translates to keeping a dual set of symmetric addresses from top and bottom, and fetching four points at a time. The table is skipped due to simplicity of derivation.

C.3.3 Scrambling and Unscrambling of Data

Due to the inherent nature of the FFT algorithm, and depending on whether DIT or DIF is chosen, input data points or outputs will be in bit reversed order in radix-2 computation. As a result, coefficient entry into the FFT engine needs to be adjusted accordingly. Bit reversal in a radix-2 case can be generalized to "digit reversal" in any radix-r architectures. Many attempts have been made to study the feasibility of implementing this type of address generation in software, but due to the long delays of software implementations, the consensus has been to add the hardware to the system. Hardware implementation can be greatly simplified if the reordering of data is properly designed into the memory access blocks. An example of the duality between radix-2 and radix-4 data input and output orderings is given in Table C.2.

C.3.4 Twiddle Factor Generation

One of the hardware implementation considerations for an FFT engine is twiddle factor generation. Hardware requirements can be greatly reduced by exploiting the symmetry properties of the twiddle factor. As depicted in Figure C.5, only the values in region H need be stored; the rest of the required twiddle factors can be generated simply by inversion and transposition of the stored values. Table C.3 shows the twiddle factors for each point on the plane based on stored values for the real (R) and imaginary (I) values of the corresponding point in region H. Hence the minimum amount of read-only-memory required for twiddle factor storage is reduced to $N/8$.

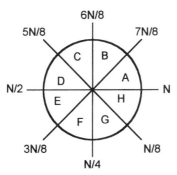

Figure C.5 Twiddle factors for *N*-Point FFT in the complex plane; only region marked as "H" is stored in ROM.

TABLE C.3 Twiddle Factor Generation

Region	Real[Twiddle Factor]	Imaginary[Twiddle Factor]
A	R	-I
B	-I	R
C	I	R
D	-R	-I
E	-R	I
F	I	-R
G	-I	-R
H	R	I

The points on this two-dimensional complex space depict values for W in counterclockwise direction, where

$$W = e^{j2\pi k/N} \quad \text{for } k = 0, \ldots, N - 1 \quad (C.10)$$

C.4   A REPRESENTATIVE FFT ENGINE IMPLEMENTATION

In this section, some of the design considerations for a representative FFT/IFFT engine are presented. The architecture is applicable to many different applications, including high-speed digital modems such as xDSL. In this implementation we are considering a 512-point FFT/IFFT engine. The processor performs a 512-point real-to-complex fast fourier transform in the FFT mode, and in the IFFT mode, it performs a 512-point complex-to-real inverse fourier transform.

C.4.1   Data Format

The data format should achieve the following objectives:

1. Maximize dynamic range of represented numbers.
2. Minimize representation error (maximize precision).

Bit Position	2U+1+V	2U−1	U	0
	Biased Exponent	Sign + Imaginary Part	Sign + Real Part	

Figure C.6   Data format bit allocation.

3. Minimize storage area.
4. Facilitate FFT computations.

Signed-magnitude number representation was chosen to facilitate some of the FFT computational steps, and floating-point number representation was chosen for highest dynamic range and reduced storage area. Each data point is represented as a complex, sign-magnitude, floating-point number with one unsigned (implicitly negative) exponent that, in order to save memory space, is shared between the real and imaginary parts.[2] The mantissa is a fully fractional quantity, and the whole complex number is always less than 1, with 1 being used to indicate overflow conditions. That is,

$$DP = (real + j * imaginary) * 2^{-exp} \qquad (C.11)$$

Figure C.6 shows the data format comprising u real bits, u imaginary bits, ν exponent bits, and one sign bit. The dynamic range of the exponent is $L = 2^{\nu}$, so that

Magnitude of smallest represented number (precision) $2^{-(L+u)}$
Magnitude of largest represented number 2^0
Range of biased exponent $[0, 2^{\nu}]$

NOTE: The same number representation must be followed on all the RAMs across the engine and all supporting system blocks. Overflow is prevented in each stage of the FFT operation by an effective scale-down performed at every stage. Furthermore, minimum exponent of each stage is tracked and it is ensured that all the exponents in the stage are scaled up by the complement of that minimum value, thus increasing the magnitude of the elements of each vector. The combination provides for smaller data sizes, prevention of overflow, and overscaling while maintaining low quantization noise levels.

C.4.2   FFT System Top-Level Architecture

The top-level system is designed as a superpipelined processing engine. Data are fetched from input files while computation is being done in the FFT processor

---

[2] In cases when it is not possible to normalize both parts of data using this scheme, there will be a slight precision loss.

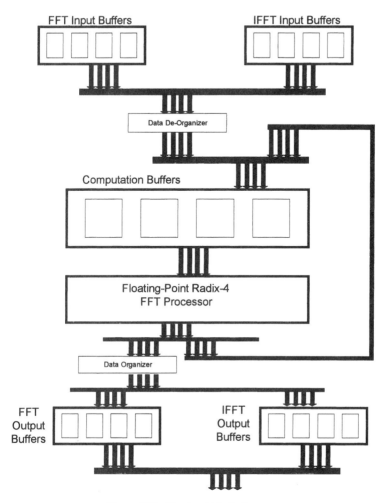

Figure C.7 Top-level block diagram.

and other data are being retired to the output stage. The processor itself is further pipelined to accommodate maximum throughput. The number of internal stages is designed to keep the output stage fully occupied as well as allowing real-time receipt of data into the input stage. Figure C.7 shows the block diagram of the FFT engine. The FFT processor is discussed further in the following sections.

C.4.3 Processor Pipeline Stages

As described earlier, for large values of N, DIT or DIF schemes are used to reduce the complex FFT operation into smaller values of N until the radix of the algorithm is reached. This process requires $\log_r N$ stages of computation.

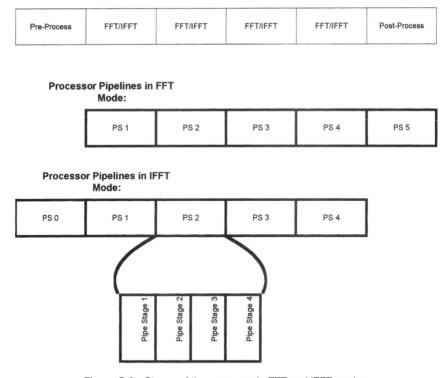

Figure C.8 Stages of the processor in FFT and IFFT modes.

Figure C.8 shows the pipeline stages for radix-4 processor for an effective 512-point FFT. The processor requires $\log_4(256) = 4$. Stages 1 through 4 refer to normal FFT stages, while stage 0 refers to preprocessing, which is active only in case of IFFT, and stage 5 refers to postprocessing, active only in FFT mode of operation.

There is a universal pipeline designed for each of the modes the FFT processor is operating in: raw_fft, pre_process and post_process. Some of the operations are modified or by-passed depending on the mode. A simplified list of operations that happen at each pipeline stage follows.

- *Stage 1*
  - Register incoming data
  - Perform sign-magnitude conversions
- *Stage 2*
  - Convert to fixed point
  - Multiplex data
  - Butterfly operations
  - Butterfly multiplications

- Perform final additions for pre/post
- Multiplex data
- *Stage 3*
  - Convert to floating for complex multiplies
  - Perform sign-magnitude conversions
  - Multiplex data
- *Stage 4*
  - Perform final adds
  - Convert to fixed point

C.4.4 Dedicated Storage Elements

Three types of storage systems are required for this type of processor:

1. Input buffers are used to receive data prior to entering the computation phase. In FFT mode they would store 512 real time-domain samples, and in IFFT mode they would store 256 complex frequency-domain taps.
2. Computation buffers store intermediate results for the various stages of in-place radix-4 FFT/IFFT computations.
3. Output buffers in FFT mode store 256 complex frequency-domain samples and in IFFT mode store 512 samples of time-domain data.

Dedicated input and output buffers for FFT and IFFT enable superpipelining requests for inverse and regular transforms. A top-level view of these storage modules is shown in Figure C.6.

REFERENCES

Many of the references here are to contributions presented to ANSI committees T1E1.4 and others in the course of their work on xDSL standards. These contributions are not "published" documents in the usual sense of the word, but they are in the public domain. They are all available on CD-ROM from Dick Bobilin, Creative Communications Consultants, P.O. Box 15189, Durham, NC 27704-0189; Tel: 908-842-6250; E-mail: dick.bobilin@t1.org. There are also a few contributions to ETSI committee TM6; these, unfortunately, are not in the public domain, so inquiries should be addressed to the ETSI secretariat.

Note: In assessing the topicality of the papers referenced here, readers should bear in mind the typical delay between writing and publication: for contributions to standards bodies (T1E1, etc.) it may be as little as three days; for ICC and Globecom it is about nine months; for IEEE Transactions it is two to three years.

[ADSLF, 1998]: ADSL Forum, TR-006, "SNMP-based ADSL line MIB," 1998.

[ANSI, 1993a]: "Digital hierarchy: layer 1 in-service digital transmission performance monitoring," ANSI Standard T1.213-1993.

[ANSI, 1993b]: "ISDN basic rate network side of NT, Layer 1," ANSI Standard, 1993.

[ANSI, 1994]: "TR-28: a technical report on high-bit rate digital subscriber lines," Committee T1-Telecommunications, Feb. 1994.

[ANSI, 1995]: "Asymmetric digital subscriber line (ADSL) metallic interface," ANSI Standard T1.413-1995.

[Antoniou, 1969]: A. Antoniou, "Realization of gyrators using op-amps, and their use in RC-active network synthesis," Proc. IEE, pp. 1838–1850, Nov. 1969.

[Armstrong, 1998]: J. Armstrong, "Polynomial cancellation coding to reduce interference due to Doppler spread," Globecom, pp. 3221–3226, Nov. 1998.

[Aslanis et al., 1992]: J. T. Aslanis, P. T. Tong, and T. N. Zogakis, "An ADSL proposal for selectable forward error correction with convolutional interleaving," T1E1.4/92-180, Aug. 1992.

[ASTM, 1994]: "Standard test methods for electrical performance of insulations and jackets for telecommunications wire and cable," American Society for Testing of Materials, ASTM D4566, 1994.

[AT&T, 1982]: Members of the Technical Staff, Bell Telephone Laboratories, *Transmission systems for communications*, AT and T Bell Laboratories, 1982.

275

[AT&T, 1983]: Bell Syst. Tech. Ref., "Digital data system: channel interface specification," PUB 62310, Sept. 1983.

[Bello, 1964]:P. A. Bello, "Characterization of randomly time-variant linear channels," IEEE Trans. Commun., Dec. 1963.

[Bengtsson et al., 1997]: D. Bengtsson, P. Deutgen, N. Grip, M. Isaksson, L. Olsson, F. Sjöberg, and H. Öhman, "Zipper performance when mixing ADSL and VDSL in terms of reach and capacity," T1E1.4/97-138, May 1997.

[Berlekamp, 1980]: E. Berlekamp, "The technology of error-correcting codes," Proc. IEEE, pp. 564–592, May 1980.

[Bingham, 1988]: J. A. C. Bingham, *The theory and practice of modem design*, Wiley, New York, 1988.

[Bingham, 1990]: – – –, "Multicarrier modulation for data transmission: an idea whose time has come," IEEE Commun. Mag., pp. 5–14, May 1990.

[Bingham,1993]: – – –, "Proposal for a symbol-synchronized scrambler," T1E1.4/93-182, Aug. 1993.

[Bingham, 1995]: – – –, "In-band digital audio radio: an update on the AT&T/Amati PAC/DMT solution," 2nd International Symposium on Digital Audio Broadcasting, pp. 270–275, March 1994.

[Bingham, 1996]: – – –, "Synchronized DMT for low-complexity VDSL," T1E1.4/96-081, Apr. 1996.

[Bingham et al., 1996a]: J. A. C. Bingham, J. M. Cioffi, and M. Mallory, "Quantifying the problem of RFI ingress in VDSL," T1E1.4/96-082, Apr. 1996.

[Bingham et al., 1996b]: J. A. C. Bingham, J. M. Cioffi, and J. S. Chow, "SDMT: a proposal for a system architecture and line code for VDSL," ETSI STC / TM6, TD-6, Stockholm, Apr. 1996.

[Bode, 1945]: H. W. Bode, *Network analysis and feedback amplifier design*, Van Nostrand, New York, 1945.

[Burrus et al., 1998]: S. Burrus, R. A. Gopinath, and H. Guo, *Introduction to wavelets and wavelet transforms: a primer*, Prentice Hall, Upper Saddle River, N.Y., 1998.

[Chang, 1996]: R. W. Chang, "Synthesis of band-limited orthogonal signals for multichannel data transmission," Bell Syst. Tech. J., pp. 1775–1796, Dec. 1996.

[Chow and Bingham, 1998]: J. S. Chow and J. A. C. Bingham, "Method and apparatus for superframe bit allocation," U.S. patent application, Jan. 1998.

[Chow et al., 1993]: J. S. Chow, J. M. Cioffi, and J. A. C. Bingham, "Equalizer training algorithms for multicarrier modulation systems," International Conference on Communications, pp. 761–765, May 1993.

[Chow et al., 1998]: J. S. Chow, J. A. C. Bingham, M. B. Flowers, and J. M. Cioffi, "Mitigating clipping and quantization effects in digital transmission systems," U.S. patents 5,623,513, Apr. 1997; 5,787,113, July 1998.

[Cioffi, 1991]: J. M. Cioffi, "A multicarrier primer," T1E1.4/91-157, Nov. 1991.

[Cioffi, 1998]: J. M. Cioffi editor, "VDSL systems requirements," T1E1.4/98-043R1.

[Cioffi et al., 1996]: J. M. Cioffi, M. Mallory, and J. A. C. Bingham, "Analog RFI cancellation with DMT," T1E1.4/96-084, Apr. 1996.

[Cioffi et al., 1999]: J. M. Cioffi, V. Oksman. J.-J. Werner, T. Pollet, P. M. P. Spruyt, J. S. Chow, and K. S. Jacobsen, IEEE Commun. Mag., pp. 72–79, Apr. 1999.

[Clark and Cain, 1981]: G. C. Clark and J. B. Cain, *Error-correction coding for digital communications*, Plenum, New York, 1981.

[Cook, 1994]: J. Cook, "Telephony transmission and splitters, passive and active," T1E1.4/94-043, Feb. 1994.

[Cravis and Crater, 1963]: H. Cravis and T. V. Crater, "Engineering of T1 carrier system repeatered lines," Bell Syst. Tech. J., pp. 431–486, Mar. 1963.

[Decker et al., 1990]: D. W. Decker, G. A. Anwyl, M. D. Dankberg, M. J. Miller, S. R. Hart, and K. A. Jaska, "Multichannel trellis encoder/decoder," U.S. patent 4,980,897, Dec. 1990.

[Doelz et al., 1957]: M. L. Doelz, E. T. Heald, and D. L. Martin, "Binary data transmission techniques for linear systems," Proc. IRE, pp. 656–661, May 1957.

[Dwight, 1961]: H. B. Dwight, *Tables of integrals and other mathematical data*, Macmillan, New York, 1961.

[ETSI, 1998]: "Very high speed digital subscriber line (VDSL); Part 1: Functional requirements," TS 101 270-1, Nov. 1998.

[Falconer and Magee, 1973]: D. D. Falconer and F. R. Magee, "Adaptive channel memory truncation for maximum likelihood sequence estimation," Bell Syst. Tech. J. pp. 1541–1563, Nov. 1973.

[Fegreus, 1986]: J. Fegreus, "Prestissimo," Digital Rev., pp. 82–87, Apr. 1986.

[Fliege, 1973]: N. Fliege, "A new class of second-order RC-active filters with two operational amplifiers," Nachrichten Technische Zeitung, pp. 279–282, 1973.

[Flowers et al., 1998]: M. B. Flowers, J. A. C. Bingham, and M. D. Agah, "Method and apparatus for randomized oversampling," U.S. patent 5,754,592, May 1998.

[Foster et al., 1997]: K. T. Foster, D. E. A. Clarke, F. Pythoud, M. Vautier, B. Capelle, L. Petrini, P. Priotti, T. Helmes, and M. Friese, "VDSL copper transport system," IEEE 7th International Workshop on Optical/Hybrid Access Networks, Atlanta, Mar. 1997.

[Freeman, 1981]: R. L. Freeman, *Telecommunication transmission handbook*, Wiley, New York, 1981.

[FSAN, 1996]: Report of the Full Service Access Networks Conference, London, June 1996.

[FSAN, 1999]: Fullservice Access Networks Working Group, "Proposal for VDSL duplexing method," ETSI TM6, 99T10R0, Feb, 1999.

[Gardner, 1979]: F. M. Gardner, *Phaselock techniques*, Wiley, New, York, 1979.

[Gatherer and Polley,1998]: A. Gatherer and M. Polley, "Controlling clipping probability in DMT transmission," Asilomar Conference on Circuits and Systems, Nov. 1997.

[Gresh, 1969]: P. A. Gresh, "Physical and transmission characteristics of customer loop plant," Bell Syst. Tech. J., pp. 3337–3385, Dec. 1969.

[Harashima and Miyakawa, 1972]: H. Harashima and H. Miyakawa, "Matched transmission technique for channels with intersymbol interference," IEEE Trans. Commun., pp. 774–780, 1972.

[Hare and Gruber,1996]: E. Hare and M. Gruber, "Operating parameters of typical HF U.S. amateur stations," T1E1.4 / 96-366, Nov. 1996.

[Heller et al., 1996]: P. N. Heller, S. D. Sandberg, and M. A. Tzannes, "DWMT equalization strategies for upstream VDSL," T1E1.4 / 96-205, July 1996.

[Hirosaki, 1981]: B. Hirosaki, "An orthogonally multiplexed QAM system using the discrete Fourier transform," IEEE Trans. Commun., pp. 982–989, July, 1981

[Hirosaki et al., 1986]: B. Hirosaki, S. Hasegawa, and A. Sabato, "Advanced group-band modem using orthogonally multiplexed QAM technique," IEEE Trans. Commun., pp. 587–592, June 1986.

[Hohhof, 1994]: K. Hohhof, "Return loss simulation results," T1E1.4/94-166, Sept. 1994.

[Holsinger, 1964]: J. L. Holsinger, "Digital communications over fixed time-continuous channels with memory," Lincoln Lab. Tech. Rep. 366, Oct. 1966.

[Honig and Messerschmitt, 1984]: M. L. Honig and D. G. Messerschmitt, *Adaptive filters: structures, algorithms, and applications*, Kluwer, Dordrecht, The Netherlands, 1984.

[Honig et al., 1990]: M. L. Honig, K. Steiglitz, and B. Gopinath, "Multichannel signal processing for data communications in the presence of crosstalk," IEEE Trans. Commun., pp. 551–558, Apr. 1990.

[Huang and Werner, 1997]: G. Huang and J.-J. Werner, "Cable characteristics," T1E1.4/ 97-169, May 1997.

[Hughes-Hartogs, 1987]: D. Hughes-Hartogs, "Ensemble modem structure for imperfect transmission media," U.S. patents 4,679,227, July 1987; 4,731,816, Mar. 1988; and 4,833,706, May 1989.

[Hunt and Chow, 1995]: R. R. Hunt and P. S. Chow, "Updating bit allocations in a multicarrier transmission system," U.S. patent 5,400,322, Mar. 1995.

[Isabelle and Lim, 1990]: H. Isabelle and J. S. Lim, "On modulated filter banks for image coding applications," IEEE Digital Signal Processing Workshop, New Paltz, N.Y., Sept. 1990.

[Isaksson et al., 1997]: M. Isaksson, D. Bengtsson, P. Deutgen, M. Sandell, F. Sjöberg, P. Ödling, and H. Öhman, "Zipper: a duplex scheme for VDSL based on DMT," T1E1.4/97-016, Feb. 1997.

[Isaksson and Mestdagh, 1998]: M. Isaksson, and D. Mestdagh, "Pulse shaping with Zipper: spectral compatibility and asynchrony," T1E1.4/98-041, Mar. 1998.

[Isaksson et al., 1998a]: M. Isaksson, R. Nilsson, F. Sjöberg, P. Ödling, D. Bengtsson, and D. Mestdagh, "Asynchronous Zipper mode," ETSI TM6 Technical Report 982t16, April 1998.

[Isaksson et al., 1998b]: M. Isaksson,, P. Deutgen, F. Sjöberg, S. K. Wilson, P. Ödling, P. Börjessen, "Zipper: a duplex scheme for VDSL based on DMT," International Conference on Communications. S29.7, June 1998.

[ITU, 1998]: "ATM layer maintenance," ITU-T Recommendation, 1998.

[Jablon, 1989]: N. K. Jablon, "Complexity of frequency-domain adaptive filtering for data modems," Asilomar Conference on Circuits and Systems, Nov. 1989.

[Jacobsen, 1996]: K.S. Jacobsen, "Discrete multitone-based communications in the reverse channel of hybrid fiber coax networks," Ph.D. thesis, Stanford University, 1996.

[Jacobsen et al., 1995]: K. S. Jacobsen, J. A. C. Bingham, and J. M. Cioffi, "Synchronized DMT for multipoint-to-point communications on HFC networks," Globecom, pp. 963–966, Nov. 1995.

[Kaden, 1959]:?. Kaden, *Wirbelstrome und Schirmung in der Nachtrichtentechnik*, Springer-Verlag, Berlin, 1959.

[Kailath, 1980]: T. Kailath, *Linear systems*, Prentice Hall, Upper Saddle River, N.J., 1980.

[Kalet, 1989]: I. Kalet, "The multitone channel," pp. 1704–1710, International Conference on Communications, 1987; also IEEE Trans. Commun., pp. 119–124, Feb. 1989.

[Kasturia and Cioffi, 1988]: S. Kasturia and J. M. Cioffi, "Vector coding with decision feedback equalization for partial response channels," pp. 0853–0857, Globecom, Dec. 1988.

[Keasler and Bitzer, 1980]: W. E. Keasler and D. L. Bitzer, "High-speed modem suitable for operaton with a switched network," U.S. patent 4,206,320, June 1980.

[Koilpillai and Vaidyanathan, 1992]: R. D. Koilpillai and P. P. Vaidyanathan, "Cosine-modulated filter banks satisfying perfect reconstruction," IEEE Trans. Signal Process., pp. 770–777, Apr. 1992.

[Kovacevic et al., 1989]: J. Kovacevic, D. J. Le Gall, and M. Veterli, "Image coding with windowed modulated filter banks," International Conference on Acoustics, Speech and Signal Processing, Glasgow, Scotland, May 1989.

[Krechmer, 1996]: K. Krechmer, "Technical standards: foundations for the future," ACM Standard View, pp. 4–8, Mar. 1996.

[Kretzmer, 1965]: E. R. Kretzmer, "Binary data transmission by partial response transmission," International Conference on Communications, pp. 451–455, June 1965.

[Kschischang et al., 1998]: F. Kschischang, A. Narula, and V. Eyuboglu, "A new approach to PAR control in DMT systems," ITU-T SG 15/Q4 Contribution NF-083, May 1998.

[Kurzweil, 1999]: J. Kurzweil, *An introduction to digital communication*, Wiley, New York, 1999.

[Lechleider, 1989]: J. W. Lechleider, "Line codes for digital subscriber lines," IEEE Commun., May, pp. 25–32, Sept. 1989.

[Lenahan, 1977]: J. Lenahan, "Theory of uniform cables: Part 1; Calculation of propagation parameters," Bell Syst. Tech. J., pp. 627–636, Apr. 1977.

[Lender, 1964]: A. Lender, "Correlative digital communication techniques," IEEE Trans. Commun. Tech., pp. 128–135, Dec. 1964.

[Lin, 1980]: S. H. Lin, "Statistical behavior of multipair crosstalk," Bell Syst. Tech. J., pp. 955–974, July/Aug. 1980.

[Lin and Costello, 1983]: S. Lin and D. J. Costello, *Error control coding: fundamentals and applications*, Prentice Hall, Upper Saddle River, N.J. 1983.

[Lindsey, 1972]: W. C. Lindsey, *Synchronization systems in communications and control*, Prentice Hall, Upper Saddle River, N.J., 1972.

[Mallat, 1989a]: S. G. Mallat, "Multiresolution approximation and wavelet orthonormal basis in L^2," Trans. Am. Math. Soc. vol. 315, pp. 69–87, 1989.

[Mallat, 1989b]: – – –, "A theory for multiresolution signal decomposition: the wavelet representation," IEEE Trans. on Pattern Recognition and Machine Intelligence, pp. 674–693, July 1989.

[Mallory, 1992]: M. P. Mallory, "Modulation method and apparatus for multicarrier data transmission," U.S. patent 5,128,964, July 1992.

[Malvar, 1992]: H. S. Malvar, *Signal processing with lapped transforms*, Artech House, Boston, 1992.

[Manhire, 1978]: L. M. Manhire, "Physical and transmission characteristics of customer loop plant," Bell Syst. Tech. J., pp. 35–59, Jan. 1978.

[Maxwell, 1996]: K. Maxwell, "Asymmetrical digital subscriber line: interim technology for the next forty years," IEEE Commun. Mag., pp. 100–106, Oct. 1996.

[Messerschmitt, 1974]: D. G. Messerschmitt, "Design of a finite impulse response for the Viterbi algorithm and decision feedback equalizer," IEEE International Conference Communications, pp. 37D.1–37D.5, June 1974.

[Mestdagh and Spruyt, 1996]: D. Mestdagh and P. Spruyt, "A method to reduce the probability of clipping in DMT-based transceivers," IEEE Trans. Commun., pp. 1234–1238, Oct. 1996.

[Müller and Huber,1997]: S. H. Müller and J. B. Huber, "A comparison of peak power reduction schemes for OFDM," Globecom, pp. 1–5, 1997; Electron. Lett., pp. 3680–3689, Feb. 1997.

[Musson, 1998]: J. Musson, "Maximum likelihood estimation of the primary parameters of twisted pair cables," ETSI TM6 981t08a0.

[Nasiri-Kenari et al., 1995]: M. Nasiri-Kenari, C. K. Rushforth, and A. Abbaszadeh, "Matched spectral-null codes with soft-decision outputs," IEEE Trans. Commun. pp. 677–680, Feb./Mar./Apr. 1995.

[Nguyen, 1992]: T. Q. Nguyen, "A quadratic constrained least-squares approach to the design of digital filter banks," International Symposium on Circuits and Systems, pp. 1344–1347, May 1992.

[Olsson et al., 1997]: L. Olsson, M. Isaksson, D. Bengtsson, P. Deutgen, F. Sjöberg, G. Ökvist, H. Öhman, P. Ödling, N. Grip, and P. Börjessen, "Influence of the Zipper duplex scheme on the receiver dynamic range," T1E1.4/97-139, May 1997.

[Pal et al., 1998]: D. Pal, J. M. Cioffi, and G. Iyengar, "A new method of channel shortening with applications to DMT systems," International Conference on Communications, pp. 763–768, June 1998.

[Paul, 1992]: C. R. Paul, *Introduction to electromagnetic compatibility*, Wiley, New York, 1992.

[Pécourt et al., 1999]: F. Pécourt, J. Hauptmann, and A. Tenen, "An integrated adaptive analog balancing hybrid for use in (A)DSL modems," International Solid State Circuits Conference, pp. 252–253, Feb. 1999.

[Pollakowski, 1995]: M. Pollakowski, "Eigenschaften symmetrischer Ortsanschlusskabel im Frequenzbereich bis 30 MHz," Der Fernmelde Ingenieur, vol. 9/10, Sept. 1995.

[Price, 1972]: R. Price, "Non-linearly feedback-equalized PAM vs capacity for noisy filter channels," International Conference on Communications, pp. 22.12–22.16, June 1972.

[Qureshi, 1985]: S. U. H. Qureshi, "Adaptive equalization," Proc. IEEE, pp. 1349–1387, Sept. 1985.

[Rezvani and Khalaj, 1998]: B. Rezvani and B. Khalaj, "Spectral compatibility technical report living list," T1E1.4/98-002, Mar. 1998.

[Riezenman, 1984]: M. Riezenman, "For the record: Wiener on the invention of the loading coil," Electronic Design, p. 11, Nov. 15, 1984.

[Rizos et al., 1994]: A. D. Rizos, J. G. Proakis, and T. Q. Nguyen, "Comparison of DFT and cosine-modulated filter banks in multicarrier modulation," International Conference on Communications, pp. 687–691, June 1994.

[Rockwell, 1998]: K. Ko, "Evaluation of interference between DSL and POTS channels for G.Lite splitterless operation," ITU SG 15, D.256, Geneva, Feb. 1998.

[Ruiz, 1989]: A. Ruiz, "Frequency-designed coded modulation for channels with intersymbol interference," Ph.D. thesis, Stanford University, Jan. 1989.

[Ruiz et al., 1992]: A. Ruiz, J. M. Cioffi, and S. Kasturia, "Discrete multitone modulation with coset coding for the spectrally shaped channel," IEEE Trans. Commun., pp. 1012–1029, June 1992.

[Saltzberg, 1967]: B. R. Saltzberg, "Performance of an efficient parallel data transmission system," IEEE Trans. Commun. Tech., pp. 805–811, Dec. 1967.

[Saltzberg, 1998]: – – –, "Comparison of single-carrier and multitone digital modulation for ADSL applications," IEEE Commun. Mag., pp. 114–121, Nov. 1998.

[Sandberg and Tzannes, 1995]: S. D. Sandberg and M. A. Tzannes, "Overlapped discrete multitone modulation for high speed copper wire communications," IEEE Jour. on Selected Areas in Commun., pp. 1571–1585, Dec. 1995.

[Sapphyre, 1998]: "Sapphyre loop qualification service," developed by Bellcore. More information may be obtained from Charles Wolozynski at chw@bellcore.com, but details will not be released until patent protection is assured.

[Schelkunoff, 1934]: M. Schelkunoff, "The electromagnetic theory of coaxial transmission lines and cylindrical shields," Bell Syst. Tech. J., pp. 532–579, Oct. 1934.

[Schmid et al., 1969]: P. E. Schmid, H. S. Dudley, and S. E. Skinner, "Frequency-domain partial-response signals for parallel data transmission," IEEE Trans. Commun. Tech., pp. 536–544, Oct. 1969.

[Shepherd et al., 1998]: S. Shepherd, J. Oriss, and S. Barton, "Asymptotic limits in peak envelope power reduction by redundant coding in orthogonal frequency division multiplex modulation," IEEE Trans. Commun., pp. 5–10, Jan. 1998.

[Slimane, 1998]: S. B. Slimane, "OFDM schemes with non-overlapping time waveforms," Vehicular Technology Conference, May 1998.

[Spruyt et al., 1996]: P. Spruyt, P. Reusens, and S. Braet, "Performance of improved DMT transceiver for VDSL," T1E1.4/96-104, Apr. 1996

[Starr et al., 1999]: T. Starr, J. M. Cioffi, and P. J. Silverman, *Understanding digital subscriber loop technology*, Prentice Hall, Upper Saddle River, N.J., 1999.

[Spruyt, 1997]: P. Spruyt, "Modulation of the pilot tone," TIE.4197-051, May 1997.

[Steffen et al., 1993]: P. Steffen, R. N. Heller, R. A. Copinath, and C. S. Burrus, "Theory of regular M-band wavelets," IEEE Trans. Signal Process., pp. 3497–3511, 1993.

[Telebit, 1990]: "A tutorial on multicarrier modulation for GSTN modems," EIA Committee TR30.1/90-0949; CCITT, Question 3/XVII, Sept. 1990.

[Tellado and Cioffi, 1998]: J. Tellado and J. M. Cioffi, "PAR reduction with minimum or zero bandwidth loss and low complexity," T1E1.4/98-173, June 1998.

[Tomlinson, 1971]: M. Tomlinson, "New automatic equalizer employing modulo arithmetic," Electron. Lett., pp. 138–139, 1971.

[Tong, 1998]: P. Tong, "Efficient address generation for convolutional interleaving using a minimal amount of memory," U.S. patent 5,764,649, June 1998.

[Tong et al., 1993]: P. T. Tong, T. N. Zogakis, and J. M. Cioffi, "Revised FEC and interleaving recommendations for DMT ADSL," T1E1.4/93-117, May 1993.

[Tzannes, 1993]: M. A. Tzannes, "System design issues for the DWMT transceiver," T1E1.4/93-100, Apr. 1993.

[Tzannes and Sandberg, 1997]: M. A. Tzannes and S. Sandberg "Multicarrier transmission system," U.S. patent 5,636,246, June 3, 1997.

[Tzannes and Tzannes, 1996]: M. A. Tzannes, and M. C. Tzannes, "Multicarrier transceiver," U.S. patent 5,497,398, Mar. 5, 1996.

[Tzannes et al., 1993]: M. A. Tzannes, M. C. Tzannes, and H. L. Resnikoff, "The DWMT: a multicarrier transceiver for ADSL using M-band wavelets," T1EI.4/93-067, Mar. 1993.

[Tzannes et al., 1994]: M. A. Tzannes, M. C. Tzannes, J. Proakis, and P. N. Heller, "DMT systems, DWMT systems and digital filter banks," International Conference on Communication, June 1994

[Unger, 1985]: J. H. W. Unger, "Near-end crosstalk model for line studies," T1D1.3/85-244, Nov. 1985

[Vahlin and Holte, 1994]: A. Vahlin and N. Holte, "Optimal finite duration pulses for OFDM," International Conference on Communications, pp. 256–262, June 1994.

[Vaidyanathan, 1992]: P. Vaidyanathan, *Multirate systems and filter banks*, Prentice Hall, Upper Saddle River, N.J. 1992.

[Valenti, 1997]: C. Valenti, "Cable crosstalk parameters and models," T1E1.4/97-302, Sept. 1997.

[Wallace and Tzannes, 1995]: M. Wallace and M. A. Tzannes, "Latency considerations in multicarrier systems," T1E1.4/95-056, June 1995.

[Wei, 1984]: L. F. Wei, "Rotationally invariant channel coding with expanded signal space," IEEE J. Sel. Areas Commun., pp. 659–668, Sept. 1984.

[Weinstein and Ebert, 1971]: S. B. Weinstein and P. M. Ebert "Data transmission by frequency-division multiplexing using the Discrete Fourier Transform," IEEE Trans. Commun. Tech., pp. 628–633, Oct. 1971.

[Werner and Nguyen,1996]: J. J. Werner and C. Nguyen, "On duality," T1E1.4/96-249, Sept. 1996.

[Zervos and Kalet, 1989]: N. A. Zervos and I. Kalet, "Optimized decision feedback equalization versus orthogonal frequency division multiplexing for high speed data transmission over the local cable network," International Conference on Communication, pp. 1080–1085, Sept. 1989.

[Zimmerman, 1997a], G. Zimmerman, "A new framework for spectral compatibility," T1E1.4/97-053, Feb. 1997.

[Zimmerman, 1997b], – – –, "Normative text for spectral compatibility evaluation," T1E1.4/97-180.

[Zimmermann, 1997c], – – –, "On the importance of crosstalk from mixed sources," T1E1.4/97-181, May 1997.

[Zogakis and Cioffi,1996]: T. N. Zogakis and J. M. Cioffi, "The effect of timing jitter on the performance of a discrete multitone system," IEEE Trans. Commun., pp. 799–808, July 1996.

[Zverev, 1967]: A. I. Zverev, *Handbook of filter synthesis*, Wiley, New York, 1967.

SELECTED BIBLIOGRAPHIES

Echo Cancellation

[EC1]: M. Ho, J. M. Cioffi, and J. A. C. Bingham, "An echo cancellation method for DMT with DSLs," T1E1.4/92-210, Dec. 1992.

[EC2]: D. C. Jones, "Reducing the complexity of a cyclic echo synthesizer for a DMT ADSL frequency-domain echo canceler," T1E1.4/93-255, Oct. 1993.

[EC3]: – – –, "Minimizing the complexity of the ATU-C echo canceler," T1E1.4/93-284, Nov. 1993.

[EC4]: J. Yang, S. Roy, and N. H. Lewis, "Data- driven echo cancellation for a multitone modulation system," IEEE Trans. Commun., pp. 2134–2144, May 1994.

[EC5]: J. M. Cioffi and J. A. C. Bingham, "A data-driven multitone echo canceler," IEEE Trans. Commun., pp. 2853–2869, Oct. 1994.

[EC6]: D. Jones, "Frequency domain echo cancellation for discrete multitone ADSL transceivers," IEEE Trans. Commun. pp. 1663–1672, Feb./Mar./Apr. 1995.

[EC7]: P. J. W. Melsa, R. C. Younce, and C. E. Rohrs, "Impulse response shortening for discrete multitone receivers," IEEE Trans. Commun., pp. 1662–1672, Dec. 1996.

RFI Suppression

[RFI1]: J. A. C. Bingham, J. M. Cioffi, and M. Mallory, "Digital RFI cancellation with SDMT," T1E1.4/96-083, Apr. 1996.

[RFI2]: J. A. C. Bingham, "RFI suppression in multicarrier systems," Globecom, vol. 2, pp. 1026–1030, Nov. 1996.

[RFI3]: B. Wiese and J. A. C. Bingham, "Digital radio frequency cancellation for DMT VDSL," T1E1.4/97-460, Dec. 1997.

[RFI4]: F. Sjöberg, R. Nilsson, N. Grip, P. O. Börjesson, S. K. Wilson, and P. Ördling, "Digital RFI suppression in DMT-based VDSL systems," Proceedings of the International Conference Telecommunications, vol. 2, pp. 189–193, Chalkidiki, Greece, June 1998.

[RFI5]: D. Pazaitis, J. Maris, S. Vernalde, M. Engels, and I. Bolsens, "Equalization and radio frequency interference cancellation in VDSL receivers," Globecom S133.7, Nov. 1998.

OFDM

[OFDM1]: ETSI, "Digital audio broadcasting to mobile and fixed receivers" European Telecommunications Standard ETS 300 401, Feb. 1995.

[OFDM2]: ETSI, "Digital video broadcasting: framing structure, channel coding, and modulation for digital terrestrial television," European Telecommunications Standard ETS 300 744, Mar. 1997.

[OFDM3]: W. Y. Zou and Y. Wu, "COFDM: an overview," IEEE Trans. on Broadcasting, pp. 1–8, Mar. 1995.

[OFDM4]: L. J. Cimini, "Analysis and simulation of a digital mobile channel using orthogonal frequency multiplexing," IEEE Trans. Commun. pp. 665–675, July 1985.

[OFDM5]: B. Le Floch, M. Alard, and C. Berrou, "Coded orthogonal frequency division multiplexing," Proc. IEEE, pp. 992–996, June 1995.

[OFDM6]: T. Pollet and M. Moeneclaey, "Synchronizability of OFDM signals," Globecom, pp. 2054–2058, Nov. 1995.

[OFDM7]: T. Pollet, M. van Bladel, and M. Moeneclaey, "BER sensitivity of OFDM systems to carrier frequency offset and Wiener phase noise," IEEE Trans. Commun., pp. 191–193, Feb./Mar./Apr., 1995.

[OFDM8]: P. H. Moose, "A technique for orthogonal frequency division multiplexing frequency offset correction," IEEE Trans. Commun., pp. 2908–2914, Oct. 1994.

Filter Approach to MCM

[Filt1]: G. A. Franco and G. Lachs, "An orthogonal coding technique for communications," IRE Intl. Conv. Rec., pp. 126–133, June 1961.

[Filt2]: R. W. Chang, "High-speed multichannel data transmission with band-limited orthogonal signals," Bell Syst. Tech. J., pp. 1775–1796, Dec. 1966.

[Filt3]: M. S. Zimmerman and A. L. Kirsch, "The AN/GSC-10 variable rate data modem for HF radio," IEEE Trans. Commun. Tech., pp. 197–204, Apr. 1967.

[Filt4]: B. R. Saltzberg, "Performance of an efficient parallel data transmission system," IEEE Trans. Commun., pp. 805–811, Dec. 1967.

[Filt5]: G. Cariolara, D. Tognetti, and G. Vannucchi, "A new architecture for orthogonally multiplexed QAM systems," Globecom, pp. 1985–1989, Nov. 1993.

[Filt6]: G. Cariolara, A. Constanzi, G. Micchieletto, and L. Vangelista, "The design of an OFDM modem: an alternative approach," Proceedings of the International Workshop on HDTV, Torino, Italy, Nov. 1994.

[Filt7]: G. Cariolara and F. Vagliani, "An OFDM scheme with half complexity," IEEE J. on Sel. Areas Commun., pp. 1586–1599, Dec. 1995.

FFT Implementation

[FFT1]: J. W. Cooley and J. W. Tukey, "An algorithm for the machine computation of complex Fourier series," Math. Comput., vol. 19, pp. 297–301, 1965.

[FFT2]: A. V. Oppenheim and R. W. Schafer, *Digital signal processing*, Prentice Hall, Upper Saddle River, N.J., 1975.

[FFT3]: J. Eldon, "A 22-bit floating-point registered arithmetic logic unit," International Conference on Acoustics, Speech and Signal Processing, pp. 943–946, Boston, 1983.

[FFT4]: G. E. Winer and R. R. Yamashita, "A single-board floating-point signal processor," International Conference on Acoustics, Speech and Signal Processing, pp. 947–950, Boston, 1983.

[FFT5]: E. O. Nwachukwu, "Address generation in an array processor," IEEE Trans. Comput., pp. 170–173, Feb. 1985.

[FFT6]: S. Naffziger, "A sub-nanosecond 0.5μm 64b adder design," International Solid State Circuits Conference, Feb. 1996.

INDEX